Conservation of Caribbean Island Herpetofaunas

Volume 1: Conservation Biology and the Wider Caribbean

Conservation of Caribbean Island Herpetofaunas

Volume 1: Conservation Biology and the Wider Caribbean

Edited by

Adrian Hailey
Byron S. Wilson
Julia A. Horrocks

BRILL

LEIDEN • BOSTON
2011

Cover photo: *Pristimantis urichi*, endemic to Trinidad and Tobago, from Trinidad (photo by A. Hailey). IUCN Status: Endangered.

This book is printed on acid-free paper.

Library of Congress Cataloging-in-Publication Data

Conservation of Caribbean island herpetofaunas / edited by Adrian Hailey, Byron S. Wilson, Julia A. Horrocks.
p. cm.
Includes bibliographical references and index.
ISBN 978-90-04-19410-6 (hardback : alk. paper) – ISBN 978-90-04-18395-7 (vol. 1 : alk. paper) – ISBN 978-90-04-19408-3 (vol. 2 : alk. paper) 1. Amphibians–Conservation–Caribbean Area. 2. Reptiles–Conservation–Caribbean Area. I. Hailey, Adrian. II. Wilson, Byron S. III. Horrocks, Julia. IV. Title.

QL656.A1C66 2011
597.8–dc22

2011008986

ISBN 978-90-04-18395-7 (volume 1)
ISBN 978-90-04-19408-3 (volume 2)
ISBN 978-90-04-19410-6 (set)

Contents

Volume 2

Preface

A set of reviews on the Conservation of Caribbean Island Herpetofaunas was originally proposed, in August 2005, as a special issue of the journal *Applied Herpetology* under the guest editorship of herpetologists at the three campuses of the University of the West Indies, with the encouragement of the journal's Editor-in-Chief, Mike Tyler. What proved to be the first part of a series was published in August 2006, and as the project expanded five parts were issued up to April 2009. Those included several non-review papers and others that were independently submitted to the journal and edited by Mike Tyler and then incorporated into the series (contents are listed at *http://www.ahailey.f9.co.uk/appliedherpetology/cariherp.htm*). *Applied Herpetology* ceased publication at the end of 2009, still leaving several invited papers for the Caribbean series in various stages of completion. Brill, the publishers of the journal, agreed to reprint (with addenda giving updated information) the review papers from *Applied Herpetology* and the subsequently completed papers in book form. The first volume includes the non-geographic reviews (on taxa or issues), plus three reviews on territories that form part of the wider Caribbean region but are not biogeographically part of the West Indies. The second volume includes geographic reviews on the West Indies proper. Of the 26 papers in the two volumes, about a third (9) are new and the remaining 17 were previously published in *Applied Herpetology*. Apart from the Introduction, all the papers have been through the same assessment procedure of the original journal, involving at least two independent anonymous referees (to whom we are grateful for helping to maintain the quality of the series), and one or more stages of revision before acceptance. The editor responsible and the original acceptance date are indicated for each paper.

Our aim was to provide a comprehensive set of articles detailing conservation concerns and actions on the amphibians and reptiles (including sea turtles) of the territories (islands or island groups) of the wider Caribbean, a region of notable biodiversity. The participation of Caribbean-based authors was especially encouraged. It was expected that the level of detail would vary among papers, depending on island size and herpetological richness, and the amount of published work available. Subjects for consideration included the general ecology of the islands; their human history, especially relating to development; major environmental problems; conservation legislation; protected areas relevant to herpetofauna; conservation status of the herpetofauna; notable species or groups; conservation programmes or ecotourism involving amphibians and reptiles; introduced herpetofauna; and recommendations. The authors have, we believe, provided an exemplary set of papers describing the conservation situation of herpetofauna in the wider Caribbean, and in many cases charted the way ahead for the territories covered. We were unable to source accounts for the remaining territories (although almost 50 contributions were invited), but the gaps will hopefully stimulate new synthesis for those areas. It was economically possible to include only monochrome photographs in either the journal or the book, but colour versions of many of those are available at the web site address above. We thank Dr Margarita C. Cuevas Gozalo, the project editor at Brill for these volumes (and previously for *Applied Herpetology*), for her consistent help.

Adrian Hailey, Byron S. Wilson, Julia A. Horrocks
The University of the West Indies

Introduction

Conservation of insular herpetofaunas in the West Indies

Byron S. Wilson[1], Julia A. Horrocks[2], Adrian Hailey[3]

[1] Department of Life Sciences, University of the West Indies, Mona, Kingston 7, Jamaica. Corresponding author; e-mail: byron.wilson@uwimona.edu.jm
[2] Department of Biological and Chemical Sciences, University of the West Indies, Cave Hill, Barbados
[3] Department of Life Sciences, University of the West Indies, St. Augustine, Trinidad and Tobago

Key words: Amphibian; biodiversity; Caribbean; conservation; habitat loss; Haiti; invasive species; reptile; sea turtles; West Indies.

Introduction

This special issue series of *Applied Herpetology* assesses the status of amphibian and reptile conservation efforts in the insular West Indies. Most of the invited contributions will cover single islands or island groups, but some papers will have a thematic focus on a taxon (e.g. snakes, amphibians, sea turtles) or conservation issue (e.g., invasive species, recent extinctions). Our area of geographic coverage is the insular western Atlantic tropical coral province, a region containing all islands within the Caribbean Sea, plus the Bahamas, the Turks and Caicos islands, and Bermuda, with the aim of discussing the herpetological conservation problems of small island developing states (Griffith and Ashe, 1993; Oldfield and Sheppard, 1997; Pelling and Uitto, 2001) within this area. These islands share not only the environmental problems associated with small (or relatively small) size, but also a history (and sometimes a present) of external political control, and high population densities of largely or entirely non-indigenous peoples. They also have a similar faunal base, indeed increasingly so with species introductions – even the Bermuda islands now have more than half of their terrestrial herpetofauna being of Antillean origin (Bacon, Gray and Kitson, this volume). There is no single term for the islands in this region, although they fall within the Wider Caribbean as defined by WIDECAST (see Sea turtles, below). We will refer to this area simply as the West Indies or Caribbean, although it extends beyond the Caribbean Sea itself and does not correspond with the West Indies as used in biogeography (which excludes Trinidad and Tobago, and Aruba, Curaçao and Bonaire, apart from others mentioned above) or politics (which includes the Guianas in northern South America: Guyana, French Guiana, and Suriname). Other authors in this series may use the terms West Indies or Caribbean in more or less inclusive ways, depending on their subject matter, so these terms should always be taken in context.

Most of these islands, and certainly all of the major ones, have long histories of herpetological exploration and discovery, and even longer histories of human-mediated environmental degradation. Collectively, they constitute a major biodiversity hotspot – a region rich in endemic species that are threatened with extinction (Meyers et al., 2000; IUCN, 2004; Ricketts et al., 2005; Smith et al., 2005). Indeed, the over-riding theme of these volumes is that West Indian herpetofaunas have been, and will likely continue to be, subjected to extirpations and extinctions as a result of human activities – especially continued loss and degradation of habitats. And while some West Indian species have been able to sustain substantial human-induced habitat change, the majority of species have been negatively affected by habitat disturbance. Make no mistake about it, human-induced habitat degradation has, on the whole, had disastrous consequences for West Indian amphibians and reptiles. If current rates of habitat conversion persist, even in the absence of the human population growth projected, we are likely to witness many more extirpations and extinctions in the coming decades.

Encouragingly, however, and as the contributions to these volumes indicate, humans are also exerting a beneficial, albeit corrective, influence through directed conservation efforts. In at least a few notable examples, human interventions have perhaps even rescued some species from what otherwise would have been extirpation or extinction (e.g., Alberts, 2000), while other threatened species appear to be recovering (e.g., Dutton et al., 2005). Unfortunately, such successful conservation efforts have been rare. Moreover, not a single West Indian species listed as threatened on the IUCN Red List has been re-classified to a reduced threat status; rather, a large number of species have been either elevated to a higher threat category or have been added to the threatened list with each revision.

Globally, the frequency of amphibian and reptile declines and extirpations has increased dramatically over the last several decades (Gibbons et al., 2000; Stuart et al., 2004). The aptly termed "global amphibian crisis" has been the subject of major international concern: nearly a third (32.5%) of all amphibian species are threatened with extinction – making them the most threatened vertebrate class (Stuart et al., 2004; Young et al., 2004). Birds and mammals, the only other vertebrate classes to have been assessed on a global scale, contain only 12% and 23% threatened species, respectively (Stuart et al., 2004). As a region, the West

Indies contains the most imperiled amphibian fauna in the world (Young et al., 2004). Reptiles are now being globally assessed using IUCN Red List criteria, and will likely show threat levels that also exceed those of birds and mammals. And within the West Indies, reptiles, like amphibians, will probably show levels of endangerment that are at or near the highest worldwide.

Because amphibians and reptiles are often the most abundant vertebrates on islands in the West Indies, their collective, extreme level of endangerment must certainly have important implications for the future of the island ecosystems in which they occur. Accordingly, amphibian and reptile conservation is a pursuit that is tantamount to conserving the integrity of these fragile insular ecosystems. Because of their highly permeable skin, and often dual, aquatic-terrestrial life styles, amphibians have rightly been hailed as superlative "bioindicators" of environmental health (Young et al., 2004). And the indications are not good: 84% of West Indian amphibians are now threatened with extinction (Young et al., 2004). Terrestrial reptiles can similarly be used as indicators of overall environmental health, because they occur in similar habitats, and are sensitive to the same perturbations as amphibians (Gibbons et al., 2000). Accordingly, if West Indian amphibians and reptiles are in widespread decline – and they most assuredly are – then something must clearly be amiss within West Indian environments.

Our guiding vision has been to provide detailed, up-to-date information for all the major West Indian islands, and to present topical papers of special relevance to their herpetological conservation, as far as possible including authors based in the region. Our ultimate goal is to provide a comprehensive set of papers that will inform, guide, and inspire future herpetological conservation efforts.

Biogeographical and Political Considerations

The islands considered here have a variety of geological origins (Pindell et al., 2005; S. Mitchell, pers. comm.). Some represent active volcanic islands on the eastern margin of the Caribbean Tectonic Plate (e.g., the volcanic islands of the Lesser Antilles), one is an accretionary wedge (Barbados), while others (the Greater Antilles, the Netherlands Antilles and Tobago) represent uplifted and deformed extinct volcanic islands (and later sedimentary successions) that formed on the margin of the Caribbean Plate. The Bahamas and Turks and Caicos (together with the northern margin of Cuba) are emergent portions of the Bahamas Carbonate Platform, part of the North American Plate, whereas Trinidad is the deformed and uplifted northern margin of the South American Plate. Bermuda has a different origin; it has a sedimentary cover on top of a sea mount, and is comparable to Pacific Atolls.

As the sole continental island exception, Trinidad, as expected, contains a diverse and substantial subset of the South American fauna, and relatively few endemic species. Most other West Indian islands contain lower diversity but often boast high percentages of endemic species. Jamaica, for example, has a native amphibian fauna that contains only 21 described species – but all of them are endemic to the island. The smaller islands predictably contain fewer species, but endemic representation is often high (e.g., the Lesser Antilles). Overall, the West Indies contain a diversity of island and habitat types, which is reflected in the region boasting one of the most diverse herpetofaunas in the world, certainly on a per unit area basis.

Overlain on this geographical mosaic is a political domain that is even more kaleidoscopic. The West Indies, as we have defined it, contains multiple islands that are administered by 18 governments (table 1). Some large single islands (e.g., Cuba) are governed by a single political entity, whereas other islands, even some relatively small ones, are administered by two separate nations (e.g., St. Maarten). Political independence was attained relatively recently for many islands; others exist as protectorates; still others are administered by governments linked to their colonial past, or remain part of the metropolitan country.

Despite significant differences in current political structure and past colonization history, West Indian islands generally share four characteristics: A long history of human colonization; high population densities; high population growth rates; and low per-capita incomes (table 1). Many of these islands also share a long history of inter-island trade, including the movement of exotic plantation crops and non-native animals. These factors have conspired to threaten amphibian and reptile populations through habitat loss, habitat degradation, over-exploitation, and the introduction of invasive species. To be sure, the West Indies has been the site of a disproportionate number of historical extinctions (Corke, 1992; Henderson, 1992), and presently features a distressingly long list of globally threatened species (IUCN, 2004).

Taxonomic Considerations

Another general observation concerning West Indian herpetofaunas is that conservation attention has been uneven with respect to taxonomy. As for other animal taxa, and for other geographical regions, herpetological conservation efforts in the West Indies have generally focused on large, conspicuous (= "charismatic") species – often those for which a commercial market exists. For example, the first legislation enacted specifically for the protection of an amphibian or reptile was an act of the Bermuda Assembly, passed in 1620, to protect juvenile sea turtles. Similar laws restricting the harvest of small sea turtles and/or protecting nesting turtles and their eggs, either directly or through the use of closed seasons, were widely enacted throughout the West Indies, some as early as the beginning of the last century (e.g., the Fisheries Regulation Act of 1904 in Barbados). Notably, these early legislative efforts were geared toward the maintenance of harvestable stocks, rather than the maintenance of biodiversity. Even today, when all sea turtle species are threatened, their protection often still falls under fisheries legislation.

More recently, a plethora of laws conferring legal protection to terrestrial wildlife species in the West Indies have appeared. In addition to legislation covering selected, "sensitive" species (e.g., *Cyclura* species), general wildlife laws now afford

Table 1. Political, population, and economic data for territories included in the Conservation of Caribbean Island Herpetofaunas series.

Territory[1]	Political Status[2]	Land Area (km^2)	Population Density (km^{-2})	Population Growth (% yr^{-1})	Urban Population (%)	GDP per capita (US$ PPP)[3]	Forest Area (%)[4]
Anguilla	UK OT	102	132	1.57	–	7,500	–
Antigua and Barbuda	Independent	443	156	0.55	37.8	11,000	20.5
Aruba	Netherlands	193	372	0.44	51.5	21,800	–
Bahamas, The	Independent	10,070	30	0.64	89.4	20,200	84.1
Barbados	Independent	431	649	0.37	51.7	17,000	4.7
Bermuda	UK OT	53	1230	0.61	100.0	69,900	–
British Virgin Islands	UK OT	153	151	1.97	–	38,500	–
Cayman Islands	UK OT	262	173	2.56	100.0	32,300	50.0
Cuba	Independent	110,860	103	0.31	75.9	3,500	21.4
Dominica	Independent	754	91	−0.08	72.0	5,500	61.3
Dominican Republic	Independent	48,380	190	1.47	67.1	7,000	28.4
Grenada	Independent	344	261	0.26	39.6	5,000	14.7
Guadeloupe	OD France	1706	265	0.88	–	7,900	–
Haiti	Independent	27,560	301	2.30	37.6	1,700	3.2
Jamaica	Independent	10,831	255	0.80	57.6	4,400	30.0
Martinique	OD France	1,060	411	0.72	–	14,400	–
Montserrat	UK OT	102	93	1.05	–	3,400	–
Netherlands Antilles	Netherlands	960	231	0.79	69.7	16,000	1.3
Puerto Rico	US TCS	8,870	443	0.40	76.3	18,600	25.8
St. Kitts and Nevis	Independent	261	150	0.50	34.7	8,800	11.1
St. Lucia	Independent	606	278	1.29	38.6	5,400	14.8
St. Vincent and the Grenadines	Independent	389	303	0.26	58.1	2,900	15.4
Trinidad and Tobago	Independent	5128	208	−0.87	75.3	16,700	50.5
Turks and Caicos Islands	UK OT	430	49	2.82	–	11,500	–
Venezuelan Antilles	Venezuela	–	–	–	–	–	–
Virgin Islands (U.S.)	US T	346	314	−0.12	47.3	14,500	41.2

[1]Data from World Bank (2005) and Central Intelligence Agency (2006).
[2]Political status: Overseas Department of France; United Kingdom Overseas Territory; United States Territory; United States Territory with Commonwealth Status.
[3]Gross Domestic Product shown as United States Purchasing Power Parity.
[4]Forest area includes natural and planted trees.

protection to many West Indian amphibians and reptiles. However, enforcement of these laws has typically been weak or non-existent.

Threats and Factors Affecting Vulnerability

The factors endangering West Indian herpetofaunas are largely obvious and typical of those that impact other insular faunas. Historically, some species were eliminated by early (i.e., pre-Columbian) settlers such as the Taino Indians (Case et al., 1992). Without exception, these were large, edible, and terrestrial species such as lizards of the genus *Cyclura*. A more substantial wave of extinctions was to accompany European colonization. A combination of large-scale habitat conversion for agriculture and human settlements, and the intentional or unintentional introduction of mammals such as rats, cats, and mongooses, has resulted in a wave of extirpations and extinctions over the past 500 years (Case and Bolger, 1991; Corke, 1992; Henderson, 1992).

General ecological characteristics associated with past extinction and present endangerment include restricted range, small population size, sensitivity to habitat alteration, and vulnerability to invasive species (Gibbons et al., 2000). Other species-specific characteristics associated with demographic vulnerability include large body size, diurnality, terrestriality, and the potential for commercial exploitation (Henderson, 1992; Gibbons et al., 2000). Indeed, vulnerability to anthropogenic influence has been highly variable among West Indian amphibians and reptiles. Many species appear to thrive in human-altered habitats, and have apparently not suffered population declines (e.g., many *Anolis* species). Other species appear to have suffered rapid and catastrophic demographic impacts owing to human influence.

The human population problem: Too many people and too little land

The ultimate cause of the current global extinction crisis is the explosive growth of the human population, and this is nowhere more obvious than in the West Indies. For example, the human population in the British Virgin Islands has doubled in the last three decades, and is expected to double again in the next 35 years (Perry and Gerber, in volume 2). Barbados is ranked as one of the top 10 most densely populated sovereign states and, not surprisingly, has lost all but small patches of its primary forest. The habitat loss that necessarily attends population expansion has been an overriding threat factor in the West Indies.

In addition to the direct loss of individuals, the range and population size contraction resulting from habitat reduction will render already threatened species at even greater risk of future extinction from other causes. The loss of habitat will also render many potential reintroduction programmes unrealistic, because sufficient habitat will not be available to support re-introduced populations (e.g., the Puerto Rican toad on Anegada; Perry and Gerber, volume 2). In short, habitat loss is the single greatest threat to the persistence of plant and animal life on earth, and the problem is nowhere more acute than on islands such as those under consideration here.

Other direct human impacts: Over-exploitation, accidents, and persecution

Human over-exploitation of edible species has been a major factor in the West Indies, but is less so today, primarily because a number of sensitive taxa have already been driven to extinction. Overall, sea turtle populations in the West Indies have been greatly reduced in size, primarily as a result of over-exploitation; and this, in spite of the fact that extinction risk resulting from unsustainable harvesting was recognized hundreds of years ago (e.g., the Act of Bermuda Assembly 1620). Additional human impacts on reptile and amphibian populations include road kills, outright persecution (e.g., snakes), and collecting for the pet trade (Dodd, 1993; Tolson and Henderson, this volume). The magnitude of such impacts will obviously only increase as a result of expanding human populations.

Invasive species

The influence of introduced mammalian predators has been the most conspicuous and highly publicized invasive species issue, and this attention is no doubt warranted (Case and Bolger, 1991). The devastating impact of mammalian herbivores such as goats is also well known; such species not only threaten insular floras but also exert indirect effects on animal populations through the alteration of habitat structure and prey bases (Henderson, 1992; Tolson and Henderson, this volume). Wild pigs consume diverse animal species, represent a severe threat to reptile eggs (e.g., *Cyclura* species, sea turtles), endanger native plants and plant communities, and, like goats, indirectly impact amphibian and reptile populations by altering habitats and prey bases (Alberts, 2000; Tolson and Henderson, this volume).

Of course, non-mammalian invasives have also negatively influenced West Indian amphibian and reptile populations. For example, invasive invertebrates such as the fire ant (*Solenopsis invicta*) are known to affect amphibian and reptile populations (e.g., Allen et al., 1997), though their impact in the West Indies has been largely unstudied. And it is not just invasive predators and competitors that can adversely affect amphibian and reptile populations; invasive plants, like invasive herbivores, can also alter habitats and prey bases. The impact of invasive plants on the West Indian herpetofauna has not gone unnoticed, but has received little research or conservation attention. Many habitats in the West Indies are now dominated by non-native plant species, and this has obviously affected the distribution and abundance of amphibian and reptile populations.

Environmental pollution

Pollutants (e.g., metals, pesticides, herbicides, nitrates, and phosphates) also impact amphibian and reptile populations through both direct and indirect pathways

(Gibbons et al., 2000). However, studies addressing the impacts of pollutants on West Indian amphibian and reptile species, and on tropical faunas generally, have been rare. But the often lax laws and weak enforcement typical of many developing island states suggest that the impact of environmental pollutants should rank as a high conservation priority – perhaps especially for amphibians, but also for reptiles (Gibbons et al., 2000). The West Indies has a long history of extensive plantation agriculture, with intensive pesticide use on crops such as sugar cane and bananas, even up to modern times. Some islands have significant industrial or urban development, with associated problems of chemical pollution (e.g., Fort et al., 2006). Many islands have widespread tourism development, often with consequent near shore pollution of sea turtle foraging habitats such as coral reefs and sea grass beds.

Disease and parasitism

Disease and parasitism have recently emerged as major factors implicated in the extirpation of populations and the extinction of species, particularly amphibians. And, while it is often difficult to find the proverbial "smoking gun," the fungal parasite *Batrachochytrium dendrobatidis* appears to have been responsible for a number of amphibian extinctions, and is suspected to have been responsible for many additional extinctions and "enigmatic declines" in the neotropics (Young et al., 2004), including islands in the West Indies (e.g., Burrowes et al., 2004; Stuart et al., 2004). Declines attributable to diseases such as chytridiomycosis are especially problematic because they can apparently occur in otherwise pristine habitats (e.g., Lips, 1999), and because there are currently no preventative or curative options available. As a consequence, the only hope for saving many species may be captive management (Stuart et al., 2004). The only documented example of a disease seriously affecting a West Indian reptile is viral fibropapilloma infection in green sea turtles (Gamache & Horrocks, 1991; Eliazar et al., 2000).

Global climate change

That increased emissions of so-called "greenhouse gases" are altering the earth's climate is no longer a controversial hypothesis. Most scientists now agree that the earth is undergoing rapid human-induced climate changes that are already having detrimental consequences for the planet's biodiversity. By altering both temperatures and precipitation regimes, global climate change will alter habitats such that many areas will no longer be capable of sustaining the amphibian and reptile populations they now support. And while some currently unsuitable habitats may become inhabitable due to altered climatic regimes, the generally poor dispersal abilities of amphibians and reptiles will probably prevent them from colonizing those areas; hence, amphibians and reptiles may be especially susceptible to climate induced changes to habitat (Gibbons et al., 2000).

Other consequences of climate change will be felt at both the species and community levels. For example, species that have temperature-dependent sex determination may suffer alterations in sex ratio that could have profound implications for population viability (e.g., sea turtles, crocodilians; Gibbons et al., 2000). The spectre of rising sea levels is of major concern. Many West Indian islands are small and low in elevation; coastal lowland habitats on such islands will be devastated by even a modest rise in sea level. Not only will sea turtle nesting beaches be diminished or even obliterated (see Fish et al., 2005), but entire communities will be inundated and thus eliminated.

Additive and synergistic effects

Often, population declines and extirpations are attributable to multiple causes, acting either additively or synergistically. For example, habitat loss reduces population size, and therefore exposes populations to greater risk from other factors, such as stochastic events, and the impact of invasive predators. Synergistic effects have been strongly implicated in several sudden frog extinctions; for example, the influence of global climate change appears to have heightened vulnerability to disease and resulted in the alarmingly rapid extinction of several anuran species (Pounds et al., 1999; Kiesecker et al., 2001; Burrowes et al., 2004).

Conservation of Herpetofauna in the West Indies

As elsewhere, conservation efforts aimed at protecting amphibians and reptiles and their habitats in the West Indies have lagged far behind other areas of herpetological endeavor, and behind other areas of conservation focus. Appropriately then, the contributions in these volumes are intended not only to highlight conservation concerns and illustrate patterns and causes of herpetofaunal endangerment, but also to exemplify what has actually been done to protect amphibian and reptile populations in the region. This effort will highlight those herpetological conservation strategies that have proven effective in the islands, as well as those that have not. This should help inform future conservation initiatives by pointing to successful approaches and elucidating gaps in conservation coverage – gaps that must be filled if we are to prevent the extirpation or extinction of additional populations and species.

Sea turtles

At present, sea turtles do not enjoy adequate legislative protection in the Caribbean region. Even on islands where moratoria are now in effect (e.g., Barbados, Jamaica), or where take is regulated through size limits or closed seasons (e.g., St. Lucia, St. Vincent and the Grenadines, among others), enforcement capacity has invariably lagged behind legislative controls. Furthermore, the regulations that do exist have often been made without an understanding that the biology of these long-lived and late-maturing reptiles requires that the greatest protection be conferred on the large juveniles and breeding adults to avoid population collapse. Instead, it is the catch of smaller size classes that has historically tended to be regulated. Harmonisation of legislation to protect the life history stages of these highly migratory species within, and beyond, the political boundaries of the Caribbean basin is needed, but is complicated by the number of different parties involved.

Sea turtles are unusual animals in that they utilize both terrestrial and marine habitats. Even where protection for the animals themselves is afforded, protection of critical habitat is lacking. For instance, in Barbados, the hawksbill turtle has been fully protected since 1998 by an indefinite moratorium, but its preference for nesting on more sheltered leeward beaches has made it extremely vulnerable to nesting and foraging habitat deterioration and loss which is directly attributable to tourism development. Protection of beach habitat, in particular, is urgently required if the recovery of hawksbills in Barbados is to be achieved. In Jamaica, sea turtles have been heavily exploited for centuries and the most important sea turtle nesting beaches are now largely restricted to offshore cays. Most of these cays are legally protected, as are the turtles, but actual protection has been all but non-existent. For example, protection of turtle nesting beaches on the Portland Bight cays is an explicit goal in the management plan submitted by the organization that was subsequently delegated with management responsibility for the area. Nevertheless, several of the more important cays are inhabited, at least semi-permanently, by fishermen who harvest both nesting females and eggs, and who have also introduced dogs. That no enforcement is currently in place to stem this illegal harvest, underscores the difficulty experienced, not just in Jamaica but in the region as a whole, in actuating the "protection" in protected areas management. The result has been the generation of numerous "paper parks."

Without addressing these issues, the remnants we see today of the historically immense sea turtle populations that once characterized the West Indies, cannot recover. The energy inputs and the overall ecological impact of those historical populations must have been huge (Bjorndal and Bolten, 2003; Bjorndal and Jackson, 2003), and would have constituted an important component of energy transfer between coastal and marine ecosystems. With populations now relics of their former size, their role in ecosystem dynamics must be effectively minimal. Restoring the numbers of these "keystone" species should be a high priority conservation objective within the region.

Papers in this series will discuss sea turtle conservation on each of the islands and island groups. Nevertheless, a regional approach for conservation of sea turtles is also necessary because most of the species have significant migratory phases during their lives, and populations are widely shared between countries. This is an approach that has been advanced by WIDECAST, the Wider Caribbean Sea Turtle Conservation Network (e.g., Eckert and Abreu Grobois, 2001), which supports the efforts of governments and non-governmental organisations in the region to integrate efforts to address all the identified stressors on sea turtle populations, both those occurring within a country's jurisdiction and beyond.

Conservation Activities

Not surprisingly, habitat protection and the control or eradication of invasive predators are the most commonly recommended in situ conservation actions. Unfortunately, habitat protection has often been illusory, and efforts at invasive predator control or eradication have been infrequent, despite the significant, demonstrated threat posed by invasive species.

Habitat protection

A general tendency in the West Indies, and much to the detriment of herpetological conservation, is for conservation efforts to focus on coastal and marine habitats, rather than on interior, often more mountainous habitats. The latter areas are frequently viewed by governments as less valuable land with minimal use for development purposes. This is partly because the destruction and degradation of many natural forests in the West Indies had already occurred in the early colonial era, with the widespread removal of the primary exploitable resource (i.e., timber). Perhaps as a result of this perception, legislation to protect such areas has tended to lag behind legislation protecting coastal areas. Monitoring and enforcement objectives are also even more difficult to achieve in interior areas, which helps to explain why such efforts have tended to be inadequate or even nonexistent.

The importance of preserving inland habitats notwithstanding, it was the coastal lowland areas that were ravaged first, and most thoroughly, and whose fauna is therefore perhaps at the greatest risk. First logged and settled, then long subjected to commercial and residential development, few natural or semi-natural coastal habitats remain in the West Indies. These remnant habitats, especially those adjacent to white sand beaches, are now under severe threat from the more recent economic mainstay of the islands: tourism. Convincing governments and private landowners to conserve beach and littoral habitats is especially challenging, given the economic rewards associated with tourism. An especially disturbing pattern is the courting of foreign investors to construct large hotels on what few beaches remain. Protecting these remaining beaches, and hence protecting remaining sea turtle nesting habitats, is a conservation battle that will be hard to win. Eco-tourism development is one possible avenue that has worked quite successfully for conservation of sea turtles and their nesting habitat in this region (Troeng and Drews, 2004).

Invasive species interventions

Most previous anti-invasive species efforts have focused on the eradication of mammals on small, offshore islets; control efforts on the larger main islands have been considerably less frequent. This is no surprise, because eradication is a feasible strategy on small islands, and the conservation impact of such actions has been convincingly demonstrated. Moreover, funding agencies and conservation organizations are understandably enamored with small island eradications because of the potential for tremendous and long-lasting conservation impact. The paucity of control efforts on larger islands is no doubt attributable to the constant inputs of labor and capital that are required, and the absence of a convenient or even foreseeable endpoint to the intervention activity. So, while the wisdom and efficacy of small island eradications is unquestioned, more effort should be directed toward control efforts in high diversity, inland habitats, especially on larger islands such as those in the Greater Antilles – particularly when small, off-shore islets are not available to serve as refuges that can be rendered invasive species-free and managed with relative ease.

Recommendations

At the outset of this project some general recommendations can be made for herpetological conservation in the West Indies:

1. Review the adequacy of current legislative protection for herpetofaunas and their habitats, and propose new regulations where needed.
2. Increase the number and extent of protected areas, and increase the effectiveness of management activities – especially the enforcement of laws or regulations prohibiting habitat degradation (e.g., tree cutting) and the illegal take of threatened species.
3. Increase knowledge of the basic ecology of amphibian and reptile species. Schwartz and Henderson (1991) noted that very little was known of the ecology of 95% of the West Indian herpetofauna. Although some studies have been conducted in the past 15 years, Schwartz and Henderson's assessment is still accurate today. However, given the number of factors that are known or considered to be certain sources of threat, obvious conservation activities need not await detailed natural history information about particular species.
4. Increase taxonomic knowledge of amphibian and reptile species. Clearly, we cannot protect what we do not know exists. The West Indies continues to be the site of new species discoveries, and this work should obviously continue. Because most West Indian amphibian and reptile taxonomists reside outside the region, state agencies that make decisions about collecting permits are encouraged to grant them to legitimate scientists. The importance of reference collections cannot be underestimated; for example, examination of museum material has shed light on the disappearance of *Eleutherodactylus karlschmidti* from Puerto Rico: the chytrid fungus was detected in specimens collected in 1976 (Burrowes et al., 2004).
5. Initiate long-term monitoring efforts for as many species as possible. Long-term information on population trends is the only way to distinguish between natural fluctuations and unidirectional, human-induced declines. Few such monitoring programmes currently exist in the West Indies and these are primarily focused on sea turtles (e.g. Jumby Bay Hawksbill Project, An-

tigua). The importance of encouraging additional population studies cannot be overemphasized.

6. Increase public awareness and education regarding the important and beneficial roles played by native amphibians and reptiles. These taxa do not currently enjoy widespread appreciation in the West Indies; most species are accorded benign neglect, whereas others are persecuted out of ignorance (e.g., non-venomous snakes, lizards of the genus *Celestus*).
7. Initiate studies aimed at controlling infectious diseases in wild populations, especially monitoring for the presence of the chytrid disease that is now impacting amphibian populations around the world.
8. Investigate options for captive propagation and maintenance when the probability of extinction in the wild is demonstrably high (e.g., cases of chytridiomycosis outbreak).
9. Initiate monitoring and research activities to discern the anticipated impacts of global climate change on West Indian herpetofaunas; such efforts should be designed to identify species and habitats that are most likely to be affected rapidly and disproportionately. Conservation intervention might be possible if impacts can be predicted well in advance

Haiti: A Final Warning

An example of nearly complete environmental destruction and a major extinction of associated amphibian and reptile species (see Hedges and Woods, 1993), is probably already underway in Haiti, the poorest country in the Western Hemisphere. The impacts of poverty on a fragile biota have rarely been more obvious and more devastating, than in this densely populated country. With no natural forests remaining, and very little forest cover, the country has almost certainly sustained a number of recent extinctions that have yet to be recorded by scientists. Unfortunately, political instability has made herpetological work in Haiti problematic, especially in the past several decades.

Survey work conducted in the 1980's and early 1990's identified a number of relictual habitats that still contained a high diversity of endemic species. However, those habitats amounted to small stands of forest, often no larger than a few football fields, and it appeared clear that those sites would not persist into the 21st century (S.B. Hedges, pers. comm.). The actual loss of habitats and species has not been documented, but it seems inconceivable that it has not occurred. We highlight Haiti here, not to point to a grim outlook for the future of other West Indian biotas, but rather, to let it serve as a red flag. Jamaica in particular, should take note of the Haitian paradigm: it is a country that appears to be on a similar trajectory.

Acknowledgements. We thank the Editor-in-Chief Mike Tyler for supporting this project, and all who reviewed manuscripts, shared their ideas, or provided other forms of assistance. The help of S.B. Hedges, R. Joglar, and S. Mitchell greatly improved this manuscript.

References

Alberts, A.C., ed. (2000): West Indian Iguanas: Status Survey and Conservation Action Plan. Gland, Switzerland, IUCN.

Allen, C.R., Rice, K.G., Wojcik, D.P., Percival, H.E. (1997): Effect of red imported fire ant envenomization on neonatal American alligators. J. Herpetol. **31**: 318-321.

Bjorndal, K.A., Bolten, A.B. (2003): From ghosts to key species: restoring sea turtle populations to fulfill their ecological roles. Mar. Turtle Newsl. **100**: 16-21.

Bjorndal, K.A., Jackson, J.B.C. (2003): Roles of sea turtles in marine ecosystems: reconstructing the past. In: The Biology of Sea Turtles, vol. II, p. 259-273. Lutz, P.L., Musick, J.A., Wyneken, J., Eds, Boca Raton, CRC Press.

Burrowes, P.A., Joglar, R.L, Green, D.E. (2004): Potential causes for amphibian declines in Puerto Rico. Herpetologica **60**: 141-154.

Case, T.J., Bolger, D.T. (1991): The role of introduced species in shaping the distribution and abundance of island reptiles. Evol. Ecol. **5**: 272-290.

Case, T.J., Bolger, D.T., Richman, A.D. (1992): Reptilian extinctions: The last ten thousand years. In: Conservation Biology: The Theory and Practice of Nature Preservation and Management, p. 91-125. Fiedler, P.L., Jain, S. K., Eds, New York, Chapman and Hall.

Central Intelligence Agency (2006): The World Factbook. (http://www.cia.gov/cia/publications/factbook/).

Corke, D. (1992): The status and conservation needs of the terrestrial herpetofauna of the Windward Islands (West Indies). Biol. Conserv. **62**: 47-58.

Dodd, C.K., Jr. (1993): Strategies for snake conservation. In: Snakes: Ecology and Behavior, p. 363-393. Seigel, R.A., Collins, J.T., Eds, New York, McGraw-Hill.

Dutton, D.L., Dutton, P.H., Chaloupka, M., Boulon, R.H. (2005): Increase of a Caribbean leatherback turtle *Dermochelys coriacea* nesting population linked to long-term nest protection. Biol. Cons. **126**: 186-194.

Eckert, K.L., Abreu Grobois, F.A. (Eds). (2001): Proceedings of the Regional Meeting: Marine Turtle Conservation in the Wider Caribbean Region: A Dialogue for Effective Regional Management. Kingshill, St. Croix, U.S. Virgin Islands, WIDECAST.

Eliazar, P.J., Bjorndal, K.A., Bolten, A. (2000): Early report of fibropapilloma from St. Croix, USVI. Mar. Turtle Newsl. **89**: 13.

Fish, M.R., Cote, I.M., Gill, J.A., Jones, A.P., Renshoff, S., Watkinson, A.R. (2005): Predicting the impact of sea level rise on Caribbean sea turtle nesting habitat Conserv. Biol. **19**: 482-491.

Fort, D.J., Rogers, R.L., Buzzard, B.O., Anderson, G.D., Bacon, J.P. (2006): Deformities in cane toad (*Bufo marinus*) populations in Bermuda: Part III. Microcosm-based exposure pathway assessment. Appl. Herpetol. **3**: 257-277.

Gamache, N., Horrocks, J.H. (1991): Fibropapilloma disease in green turtles, *Chelonia mydas*, around Barbados, West Indies. In: Proceedings of the 11th Annual Workshop on Sea Turtle Biology and Conservation, p. 158-160. Salmon, M., Wyneken, J., Eds, NOAA Technical Memorandum NMFS-SEFSC-302.

Gibbons, J.W., Scott, D.E., Ryan, T.J., Buhlmann, K.A., Tuberville, T.D., Metts, B.S., Greene, J.L., Mills, T., Leiden, Y., Poppy, S., Winne, C.T. (2000): The global decline of reptiles, déjà vu amphibians. Bioscience **50**: 653-666.

Griffith, M.D., Ashe, J. (1993): Sustainable development of coastal and marine areas in small island developing states: A basis for integrated coastal management. Ocean Coast. Manage. **21**: 269-284.

Hedges, S.B., Woods, C.A. (1993): Caribbean hot spot. Nature **364**: 375.

Henderson, R.W. (1992): Consequences of predator introductions and habitat destruction on amphibians and reptiles in the post-Columbus West Indies. Carib. J. Sci. **28**: 1-10.

IUCN (2004): 2004 IUCN Red List of Endangered Species. Gland, Switzerland. Accessed online at http://www.iucnredlist.org/.

Kiesecker, J.M., Blaustein, A.R., Belden, L.K. (2001): Complex causes of amphibian population declines. Nature **410**: 681-684.

Lips, K.R. (1999): Mass mortality and population declines of anurans at an upland site in western Panama. Conserv. Biol. **13**: 117-125.

Myers, N., Mittermeier, R.A., Mittermeier, C.G., de Fonseca, G.A., Ke, J. (2000): Biodiversity hotspots for conservation priorities. Nature **403**: 853-858.

Oldfield, S., Sheppard, C. (1997): Conservation of biodiversity and research needs in the UK Dependent Territories. J. Appl. Ecol. **34**: 1111-1121.

Pelling, M., Uitto, J.I. (2001): Small island developing states: Natural disaster vulnerability and global change. Envt. Hazards **3**: 49-62.

Pindell, J., Kennan, L., Maresch, W.V., Stanek, K., Draper, G., Higgs, R. (2005): Plate-kinematics and crustal dynamics of circum-Caribbean arc-continent interactions: tectonic controls on basin development in proto-Caribbean margins. In: Caribbean-South American Plate Interactions, Venezuela, p. 7-52. Lallemant, A., Sisson, V.B., Eds, Geological Society of America Special Paper 394.

Pounds, J.A., Fogden, M.P.L., Campbell, J.H. (1999): Biological response to climate change on a tropical mountain. Nature **398**: 611-615.

Powell, R., Henderson, R.W. (2005): Conservation of Lesser Antillean Reptiles. Iguana **12**: 2-17.

Ricketts, T.H., Dinerstein, E., Boucher, T., Brooks, T.M., Butchart, S.H.M., Hoffmann, M., Lamoreux, M.J.F., Morrison, J., Parr, M., Pilgrim, J.D., Rodrigues, A.S.L., Sechrest, Wallace, G.E., Berlin, Bielby, J., Burgess, N.D., Church, D.R., Cox, N., Knox, D., Loucks, C., Luck, G.W., Master, L.L., Moore, R., Naidoo, R., Ridgely, Schatz, G.E., Shire, G., Strand, H., Wettengel, W., Wikramanayake, E. (2005): Pinpointing and preventing imminent extinctions. PNAS **102**: 18497-18501.

Schwartz, A., Henderson, R.W. (1991): Amphibians and Reptiles of the West Indies: Descriptions, Distributions, and Natural History. University of Florida Press, Gainesville.

Smith, M.L., Hedges, S.B., Buck, W., Hemphill, A., Incháustegui, S., Ivie, M., Martina, D., Maunder, M., Ortech, J.F. (2005): Caribbean Islands. In: Hotspots Revisited: Earth's Biologically Richest and Most Endangered Terrestrial Ecoregions, p. 112-118. Mittermeier, R.A., Gill, P.R., Hoffman, M., Pilgrim, J., Brooks, T., Mittermeier, C.G., Lamoreaux, J., da Fonseca, G.A.B., Eds, Mexico City, CEMEX.

Stuart, S.N., Chanson, J.S., Cox, N.A., Young, B.E., Rodrigues, A.S.L., Fischman, D.L., Waller, R.W. (2004): Status and trends of amphibian declines and extinctions worldwide. Science **306**: 1783-1786.

Troeng, S., Drews, C. (2004): Money Talks: Economic Aspects of Marine Turtle Use and Conservation. WWF-International, Gland, Switzerland (http://www.panda.org).

World Bank (2005): The Little Green Data Book 2005. Washington, DC, The World Bank.

Young, B.E., Stuart, S.N., Chanson, J.S., Cox, N.A., Boucher, T.M. (2004): Disappearing Jewels: the Status of New World Amphibians. NatureServe, Arlington, Virginia.

Reprinted from Applied Herpetology **3**: 181-195 (2006).

An overview of the evolution and conservation of West Indian amphibians and reptiles

S. Blair Hedges

Department of Biology, 208 Mueller Lab, Pennsylvania State University, University Park, Pennsylvania 16802-5301, USA
e-mail: sbh1@psu.edu

Abstract. The total area of the West Indies is small, only 0.15% of Earth's land area, but the region supports 3.0% (180 species) of the world's amphibians and 6.3% (520 species) of the world's known reptiles. Nearly all species are endemic to a single island or island bank. Rates of discovery are high indicating that many more species remain to be found. Most or all of the groups arrived at different times during the Cenozoic Era (the last 65 million years) by dispersal over water, principally on flotsam from South America. This resulted in a filtered fauna that has relatively few higher-level groups. The vacant niches led to unusually large radiations (mega-radiations) of species, explaining the presence of extremes in ecology and morphology among the species. The conservation status of only the amphibians has been assessed comprehensively, and 84% of those species are considered threatened. The reptiles may be similarly threatened. Species are threatened mostly because they have small distributions and habitat destruction is continuing. Across the region only about 10% of primary forests remain, and some countries (e.g., Haiti) with high levels of faunal diversity have no original forests. Some species that have disappeared for decades are possibly extinct, while others face extinction soon from habitat destruction. Efforts to curb deforestation and to save targeted species through breeding programs are helping, but the continued survival of many West Indian amphibians and reptiles is tenuous. Measures having immediate impact, such as salaries and logistical support for park guards, are needed immediately. Greater facilitation of biotic surveys and inventories, through relaxing of permitting restrictions, also is needed to better understand the diversity that exists and to focus conservation efforts.

Key words: Antilles; biodiversity; Caribbean; deforestation; discovery; ecology; extinction; genetic resources; systematics; taxonomy.

Introduction

There are approximately 700 species of amphibians and reptiles that have been described from the West Indies (Hedges, 2006a), here defined as excluding Trinidad and Tobago which are adjacent to South America and have continental faunas. This includes approximately 180 amphibians (3.0% of all amphibian species) and 520 reptiles (6.3% of all reptile species). These proportions of the global faunas are high when it is realized that the West Indies comprises only 0.15% of the total land area of the Earth. At least another 20 new species are under study and additional species will surely be discovered. It is too soon to estimate the total number of species that exist, but it most likely exceeds 800 and could surpass 1000 if many of the taxa recognized today as subspecies are later found to be valid species, which is a current trend. Remarkably, nearly all native species are endemic to the region and rarely found on more than one island or island bank (Hedges, 2006b).

Only the amphibians have had a comprehensive assessment of their conservation status, as part of the Global Amphibian Assessment led by Conservation International and the International Union for the Conservation of Nature (Stuart et al., 2004; Young et al., 2004). That study concluded that a surprising 84% of the amphibian species are threatened, which is the highest proportion of any amphibian fauna in the world. The primary factors responsible for placing species in the high threat categories (vulnerable, endangered, and critically endangered) were small distributions and declining habitat, mainly through deforestation. These same factors are likely to affect the conservation status of West Indian reptiles when that group is assessed in the near future, and more generally explains why the West Indies is one of the hottest of biodiversity hot spots (Myers et al., 2000; Smith et al., 2005).

Currently, only a few species — mostly large reptiles — have been targeted for captive-breeding or population monitoring and other species-specific conservation measures (e.g., Bloxam and Tonge, 1995; Tolson, 1996; Vogel et al., 1996; Daltry et al., 2001; Goodman et al., 2005). These efforts are critical, especially when population numbers are perilously low as in these cases. But the vast majority of species are effectively invisible to such programs and to the conservation community in general. They are small, cryptic, and usually live in remote areas where their declines, from habitat destruction, are not being monitored. Many of these species have only been seen by their original discoverers, or perhaps a few times after that. For this majority of the West Indian herpetofauna, extinctions will occur before any captive breeding is done, and therefore attention must be focused on habitat conservation and stopping, or at least controlling, the loss of forests. At the same time, exploration and discovery efforts need to be encouraged so that the true diversity is as completely characterized as possible, both taxonomically and ecologically. The most effective conservation strategies are those based on the most accurate knowledge of a fauna.

Here I provide brief overview of the origin and evolution of the West Indian herpetofauna. This is followed by a discussion of two aspects vital to its conservation: stopping deforestation and facilitating systematic research. Certainly, other threats exist besides deforestation, such as climate change (e.g., Malcolm et al., 2006), disease (e.g., Burrowes et al., 2004), introduced predators (e.g., Iverson, 1978), introduced flora (Henderson and Powell, 2001), and human exploitation (e.g., Lescure, 1979; Powell et al., 2000). For some species, especially large reptiles, it has been claimed that introduced predators have had a greater negative impact than habitat alteration (Henderson, 2004). Although this probably is true in some instances, the point made here is that habitat destruction — in particular, deforestation — is the major and primary threat to the survival of most species.

Origin, Evolution, and Biogeography

Knowledge of the evolutionary history of West Indian amphibians and reptiles is helpful in understanding the conservation status of these animals and threats to their survival. The historical biogeography of the West Indies has been studied for more than a century and remains an active area of research. Knowledge of geologic history is crucial for any biogeographic scenario and therefore the history of the field can be neatly divided into work published before and after the general acceptance of continental drift (c.1960s). Most early workers assumed that there was no change in the position of the islands and their water gaps, thus leaving only over-water dispersal as the mechanism (Wallace, 1881; Matthew, 1915; Darlington, 1938; Simpson, 1956). An alternative viewpoint was that ancient land bridges existed between the continents and the Antilles, permitting dispersal over land (Barbour, 1916; Schuchert, 1935). However, continental drift showed that the foundations for both views were incorrect.

We now know that the Greater Antilles were once connected as a geologic unit with North and South America in the late Cretaceous (c. 60-70 million years ago), raising the possibility that the present fauna arose by "vicariance," which consisted of faunas drifting with their islands as they broke away from the continents (Rosen, 1975). Unfortunately, fossil and genetic research has been unable to identify more than a few West Indian groups that fit this model, including *Eleutherodactylus* frogs and the Cuban xantusiid lizard *Cricosaura typica* (Hedges et al., 1992; Hedges, 2006b). The great majority of groups probably arrived in the West Indies by flying, swimming, or floating on flotsam (mats of vegetation). The predominant flow of ocean currents moving east to west means that the source for almost all flotsam in the West Indies is South America, or more rarely, Africa, which agrees with the evolutionary relationships of most of the amphibians and reptiles (Hedges, 1996b).

The strongest support for over-water dispersal as the predominant mechanism has come from the taxonomic composition of the fauna, which is missing many higher-level groups such as salamanders, caecilians, most families and subfamilies of frogs, lizards, snakes, and turtles. This was noted in even the earliest accounts of West Indian biogeography (Wallace, 1881; Matthew, 1915) and has been reiterated over the years (Darlington, 1938; Simpson, 1956; Williams, 1989; Hedges, 1996a). The fossil record, although still sparse, shows this same "filtered" fauna of the Pleistocene and Recent extending back into the mid-Cenozoic. For example, there are multiple fossils of the extant genera *Anolis*, *Sphaerodactylus*, and *Eleutherodactylus* from Dominican amber (15-20 million years ago, Ma), but no indication of other groups that should have been there if there were dry land connections to the mainland in the past. Furthermore, the large radiations of species (mega-radiations) of those extant genera, obviously filling unoccupied niches, further suggests that this filtered fauna has existed for millions of years and did not just result from recent extinctions.

All of these same reasons argue against the existence of a dry land bridge (Aves Ridge) connecting South America with the Greater Antilles in the mid-Cenozoic (Holcombe and Edgar, 1990; Iturralde-Vinent and MacPhee, 1999). Such a faunal corridor would have brought a diversity of vertebrate groups to the Antilles which would have been evident at least in the fossil record (e.g., amber fossils) if not in the living fauna. Moreover, the times of divergence of the colonists from their mainland counterparts would be reflected in molecular estimates of divergence times, clustering around 35 Ma, but instead the lack of clustering supports over-water dispersal (Hedges et al., 1992; Hedges, 1996b). Also geologic evidence is silent on the existence (or not) of such as land bridge (Holcombe and Edgar, 1990; Hedges, 2001).

Niche-filling, Ecological Release, and Biological Extremes

This history of the fauna, dominated by over-water dispersal, explains why there are so few higher-level groups (families and genera) in the West Indies and an over-abundance of species. The frequent filling of vacant niches by resident groups is called "ecological release," and a side-effect is that some of the world's smallest and largest species occur in the West Indies. Among amphibians and reptiles, this includes the smallest frog *Eleutherodactylus iberia* (Estrada and Hedges, 1996), the smallest lizard *Sphaerodactylus ariasae* (Hedges and Thomas, 2001), and the smallest snake *Leptotyphlops bilineatus*. At the other end of the size spectrum, *Osteopilus crucialis* (Jamaica) and *O. vastus* are among the largest treefrogs in the New World (Trueb and Tyler, 1974), *Alsophis anomalus* is the largest xenodontine snake, and *Cyclura nubila* (Cuba) is the largest iguanid lizard. Other species nearly reach those extremes, indicating that mega-radiations and ecological release are major features of the West Indian herpetofauna.

Other unusual aspects of the ecology and behavior of individual species may be tied to ecological release as well, including live-bearing in the Puerto Rican frog *E. jasperi*, cave breeding and froglet transport in the Jamaican frog *E. cundalli* (Diesel et al., 1995), underground chamber construction in the Hispaniolan frogs *E. hypostenor*, *E. ruthae*, and *E. pelorius*, aquatic behavior in the lizards *Anolis eugenegrahami* and *A. vermiculatus*, among others. This distinctive quality of the herpetofauna, having extremes in morphology, ecology, and reproduction tied to its evolutionary history, is difficult to ignore from a conservation standpoint.

Deforestation

Most visitors to the West Indies are unaware that the current appearance of the islands bears little resemblance to what it was when Columbus arrived in 1492. Even small islands in the Bahamas were covered with thick forests, and giant, buttressed trees were abundant. Today, only a small fraction of that original forest remains, although the actual amount is unclear. The United Nations Food and Agriculture Organization (FAO) keeps track of global forest cover in individual countries (FAO, 2005; WRI, 2006). However, caution must be used in interpreting those data. Values for the percent of forest cover include introduced trees and forest plantations, the latter of which provide limited or no habitat for native amphibians and reptiles. Also, "forest" is defined as having >10% canopy, and means that as much as 90% of the trees in an area could be removed and it would still be classified as forest.

For some countries, data are available for more stringent levels of canopy extent, up to 75%, and in a few cases, for "natural forest" and "primary forest." The FAO defines natural forest as having native species of trees, but the 10% canopy cutoff value still applies (FAO, 2005). Primary forests are defined as having "no — or no visible — indications of past or present human activity," and it is this last category that is most relevant to assessing the natural habitat and conservation status of most West Indian amphibians and reptiles. Unfortunately, the FAO lists primary forest information for only three countries in our region, and for nearby Trinidad and Tobago. Comparing total forest area with primary forest area (respectively) in these four cases reveals some major differences: Dominica (61.3%, 36%), Grenada (11.8%, 2.9%), Guadeloupe (47.3%, 11.2%), and Trinidad and Tobago (44.4%, 2.7%). Dominica is considered to have one of the highest proportions of primary forest of any island in the region, in part due to its low human population density, and should be considered an exception. The other three are probably more representative, with two having less than 3% of their original primary forest remaining. Also, two of the four countries (Guadeloupe and Trinidad and Tobago) show striking differences between total and primary forest areas.

These comparisons indicate that it can be misleading to relate the extent of total forest cover to the preservation of native faunas and their habitats. The current primary forest area of the Greater Antilles is likely to be much less than the values listed for the total forest area in those countries, which are: Cuba (22.2%), Jamaica (31.4%), Haiti (4.0%), the Dominican Republic (28.4%), and Puerto Rico (45.9%) (FAO, 2005). Instead, an estimate of 10% "intact vegetation" for the region as a whole is a more realistic figure, with some countries (e.g., Haiti) having little or no remaining primary forest (Hedges and Woods, 1993; Smith et al., 2005).

After Columbus arrived, the lowland rainforest in the Greater Antilles was the first to go, and virtually none of it remains today aside from isolated trees in old plantations and a few small patches such as one near Cabezada, Guantánamo Province, Cuba (pers. obs.). These areas were quickly cleared for agriculture, and the wood was used for building materials to feed expansion of the New World colonies. Deforestation of the mountain slopes soon followed as European settlements continued to expand, and crops (e.g., coffee) were planted that required higher elevations for optimal growth. Some small regions of montane rainforest and cloud forest still exist on most of the larger islands, but the vast majority is gone.

The original forests that remain today in the West Indies occur mostly in remote areas that are difficult to access and thus are rarely seen except on foot. Of course, it is this difficulty of access that has delayed their demise. Among these are the major limestone forests in the Greater Antilles, such as the Cockpit Country of Jamaica, the Viñales region of Cuba, the Haitises of the Dominican Republic, and the karst region of northwest Puerto Rico, and forests on some of the highest and inaccessible mountain tops. But even in these areas, access roads and trails have permitted local squatters to enter the forest illegally, clear forested valleys between mogotes, and plant crops. These cleared patches can be seen on satellite photos and make up large portions of regions designated as protected areas, including the Haitises National Park and the Cockpit Country Forest Reserve. Comparison with earlier satellite and aerial photos shows that these last remaining forests are disappearing, even in so-called protected areas (e.g., Tole, 2002). In Haiti, clearing of forest and charcoaling has left a denuded landscape (fig. 1).

Much of the lush, green roadside vegetation seen along major roadways in the West Indies is introduced vegetation from the Old World such as mango, breadfruit, papaya, banana, bamboo, and various ornamentals, weeds, and shrubs. To a casual observer, this may give the false impression that the islands are heavily forested, but this introduced vegetation is usually sparse and associated with low species diversity of the herpetofauna (Henderson and Powell, 2001). Also, forest recovery in deforested areas of the West Indies can be slow, and the regenerated forest may not resemble the original forest in terms of dominant tree species and overall species diversity (Aide et al., 1995; Rivera and Aide, 1998; Rivera et al., 2000). Even secondary impacts of deforestation are now becoming known. For example, the clearing of lowland rainforest, which has already occurred throughout the West Indies, is known to alter the climate in nearby upland forests (Lawton et al., 2001).

Habitat Protection

The most important conservation measure for saving West Indian amphibians and reptiles applies to essentially the entire West Indian terrestrial biota: Stop deforestation. Although it can be stated this simply, putting a halt to deforestation is a complicated and difficult problem, not even considering the large numbers of governments, cultures, and languages involved in the region. Almost every country has made attempts to address this problem and many have help from non-governmental organizations, which in some cases even take charge of national parks. But parks and other protected areas in the West Indies afford varying degrees of habitat protection. For example, the national parks of Haiti are essentially "paper parks," because logging and clearing of forests continues unabated within park boundaries. For this reason, *the existence of protected areas should effectively be ignored in assessing the conservation status of native species, unless there is unambiguous evidence that such a designated area is truly affording protection.* Even on islands with good enforcement of tree-cutting (e.g., Puerto Rico), native

Figure 1. (A) Deforested hills north of Jacmel on the Haitian Tiburon peninsula. (B) Charcoal being removed from a pit in the northwest peninsula of Haiti and placed in bags for transport. The clearing of forests for charcoal production (cooking fuel) is a major cause of deforestation in the West Indies. Both photographs taken in July, 1991 by S.B. Hedges.

species of amphibians and reptiles are still threatened within protected areas by abundant introduced predators (e.g., black rats). Thus, there are no "pristine" forests in the West Indies.

Plans have been proposed for protecting the natural areas of the most species-rich islands (e.g., Woods et al., 1992; Hedges and Woods, 1993; Sergile and Woods, 2001) and most focus on habitat protection. However, habitat alteration and destruction continues in the West Indies despite the hard work of many local and international conservationists and funding agencies. Because extinctions are eminent in some areas, efforts should focus now on achieving immediate benefits by, for example, funding of park guards and their associated logistical needs (e.g., supplies and vehicles).

Exploration, Discovery, and Systematics

Discovery curves are graphs that show the cumulative number of described species at different points in time. For West Indian amphibians and reptiles, these curves show steeply rising slopes indicating that many more species remain to be discovered (fig. 2). However, relatively few systematists are now conducting herpetological survey work in the region. In part it is related to the small total number of taxonomists globally and their continuing decline in numbers as an outcome of cultural shifts in scientific research and funding (Raven, 2004; Wilson, 2004). Modern systematics usually involves DNA sequencing, phylogenetic analyses, morphological descriptive analyses, and biostatistics, all of which require training and most of which is expensive. This is especially a problem for local scientists in the West Indies where funding opportunities are limited. It is important that conservation agencies and funding programs recognize that such survey and inventory work is

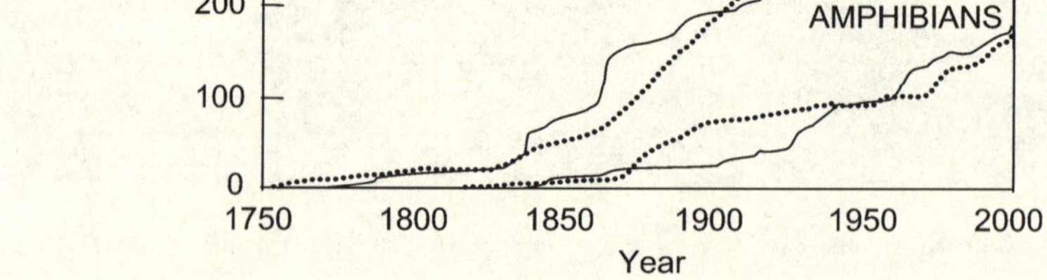

Figure 2. Discovery curves for West Indian amphibians and reptiles (continuous lines) compared with those for Malagasy amphibians and reptiles (dashed lines). The lines show cumulative numbers of described native species at each time period.

vital to conservation efforts, and that they make concerted efforts, through targeted grant funding, to facilitate such research.

A separate problem that impedes the work of systematists who conduct survey-and-inventory research in the West Indies involves restrictions imposed by countries on areas that can be visited, species and numbers of individuals that can be collected, and tissue samples that can be taken for molecular systematic research. These restrictions affect both local and foreign scientists, but can be particularly discouraging for the latter. When faced with the high costs of an expedition and the likelihood that the research will be greatly curtailed by restrictions to collecting, especially the taking of tissues for molecular systematics, many foreign scientists simply do not bother making the trip or choose to do work elsewhere, in a country that is less restrictive to scientific research. Thus, one can be certain that the small number of systematists conducting surveys and inventories in the West Indies is in part due to inhibition of research caused by these restrictions.

Biotic survey work in Cuba has been greatly curtailed during the last decade by severe restrictions on travel around the country imposed by the Cuban military. This has affected research by both Cuban and foreign scientists. If the Cuban government believes that systematists represent a security threat, a possible solution might be to assign a military escort to the expedition, funded by the expedition sponsor. Many would consider that additional expense to be a better alternative than canceling the work altogether.

It is understandable that the number of individuals to be collected should be regulated by governments, and almost all countries impose limitations. Nonetheless, consideration should be given to the fact that small animals often exist in high population densities and that scientists need to sample as broadly as possible to detect geographic variants and assess population variation. For example, densities as high as 23,600 ha^{-1} for *Anolis* (Reagan, 1992) and 52,800 ha^{-1} for *Sphaerodactylus* (Rodda et al., 2001), the two most species-rich genera of West Indian lizards, have been recorded. Scientific collecting is unlikely to affect the health of such populations, or even those with densities that are orders of magnitude smaller. The benefits of routine geographic sampling of populations are seen frequently in studies that have revealed cryptic species, such as the discovery of 16 new species of salamanders in the eastern U.S. (Highton et al., 1989). For conservation, this information is priceless.

Yet scientists are often told not to collect more than a small number of individuals (e.g., 1-5) of these abundant vertebrates, reducing the potential value of the research. This is unfortunate, because in all realistic cases that can be envisioned, *the scientist is not the threat*. The number of individuals of most species killed by clearing a single hectare of forest habitat, by the building a single house or garden, by vehicles on highways, or by a storm that blows over some trees, will far exceed the sampling effect of systematists (Hedges and Thomas, 1991).

Finally, another major problem is occurring throughout the world which is impeding conservation efforts. It is the restriction on collection of "genetic resources."

To conservation biologists, tissues samples for DNA sequencing provide invaluable data on a spectrum of ecological and evolutionary questions that can facilitate conservation efforts. For example, DNA sequences can tell us how many species exist in an area, especially with cryptic organisms, as already noted. Also, they can provide critical data about the population structure, related to population size and dispersal. The problem is actually a misunderstanding that conservation biologists will profit from these samples by finding new drugs worth millions or billions of dollars, through a procedure called "bioprospecting." Such fears are unwarranted. But to alleviate these concerns, a possible solution might be to stipulate on collecting and export permits that the samples are not to be used for drug discovery, rather than to restrict collection of the samples themselves.

Acknowledgements. I thank Robert Henderson, Robert Powell, Peter Tolson, and Byron Wilson for helpful comments on the manuscript. This work was supported by the U.S. National Science Foundation.

References

Aide, T.M., Zimmerman, J.K., Herrera, L., Rosario, M., Serrano, M. (1995): Forest recovery in abandoned tropical pastures in Puerto Rico. Forest Ecol. Manag. **77**: 77-86.

Barbour, T. (1916): Some remarks upon Matthew's "climate and evolution". Ann. NY Acad. Sci. **27**: 1-15.

Bloxam, Q.M.C., Tonge, S.J. (1995): Amphibians — suitable candidates for breeding-release programs. Biodivers. Conserv. **4**: 636-644.

Burrowes, P.A., Joglar, R.L., Green, D.E. (2004): Potential causes for amphibian declines in Puerto Rico. Herpetologica **60**: 141-154.

Daltry, J.C., Bloxam, Q., Cooper, G., Day, M.L., Hartley, J., Henry, M., Lindsay, K., Smith, B.E. (2001): Five years of conserving the 'world's rarest snake', the Antiguan racer *Alsophis antiguae*. Oryx **35**: 119-127.

Darlington, P.J. (1938): The origin of the fauna of the Greater Antilles, with discussion of dispersal of animals over water and through the air. Q. Rev. Biol. **13**: 274-300.

Diesel, R., Baurle, G., Vogel, P. (1995): Cave breeding and froglet transport — a novel pattern of anuran brood care in the Jamaican frog, *Eleutherodactylus cundalli*. Copeia **1995**: 354-360.

Estrada, A.R., Hedges, S.B. (1996): At the lower size limit in tetrapods: a new diminutive frog from Cuba (Leptodactylidae: *Eleutherodactylus*). Copeia **1996**: 852-859.

FAO (2005): Global Forest Resources Assessment 2005: Progress Towards Sustainable Forest Management. United Nations Food and Agriculture Organization, Rome.

Goodman, R.M., Echternacht, A.C., Burton, F.J. (2005): Spatial ecology of the endangered iguana, *Cyclura lewisi*, in a disturbed setting on Grand Cayman. J. Herpetol. **39**: 402-408.

Hedges, S.B. (1996a): Historical biogeography of West Indian vertebrates. Ann. Rev. Ecol. Syst. **27**: 163-196.

Hedges, S.B. (1996b): The origin of West Indian amphibians and reptiles. In: Contributions to West Indian Herpetology: A Tribute to Albert Schwartz, p. 95-127. Powell, R., Henderson, R.W., Eds, Society for the Study of Amphibians and Reptiles, Ithaca.

Hedges, S.B. (2001): Caribbean biogeography: an outline. In: Biogeography of the West Indies: Patterns and Perspectives, p. 15-33. Woods, C.A., Sergile, F.E., Eds, CRC Press, Boca Raton, Florida.

Hedges, S.B. (2006a): Caribherp: Database of West Indian Amphibians and Reptiles (http://www.caribherp.net/). Accessed 18 July 2006. Pennsylvania State University, University Park, Pennsylvania.

Hedges, S.B. (2006b): Paleogeography of the Antilles and the origin of West Indian terrestrial vertebrates. Ann. Mo. Bot. Gard. **93**: 231-244.

Hedges, S.B., Hass, C.A., Maxson, L.R. (1992): Caribbean biogeography: Molecular evidence for dispersal in West Indian terrestrial vertebrates. Proc. Natl. Acad. Sci. USA **89**: 1909-1913.

Hedges, S.B., Thomas, R. (1991): The importance of systematic research in the conservation of amphibian and reptile populations. In: Status y Distribución de los Reptiles y Anfibios de la Región de Puerto Rico. Moreno, J.A., San Juan, Eds, Departamento de Recursos Naturales de Puerto Rico, Puerto Rico.

Hedges, S.B., Thomas, R. (2001): At the lower size limit in amniotes: a new diminutive lizard from the West Indies. Caribb. J. Sci. **37**: 168-173.

Hedges, S.B., Woods, C.A. (1993): Caribbean hot spot. Nature **364**: 375.

Henderson, R.W. (2004): Lesser Antillean snake faunas: distribution, ecology, and conservation concerns. Oryx **38**: 311-320.

Henderson, R.W., Powell, R. (2001): Responses by the West Indian herpetofauna to human-influenced resources. Caribb. J. Sci. **37**: 41-54.

Highton, R., Maha, G.C., Maxson, L.R. (1989): Biochemical evolution in the slimy salamanders of the *Plethodon glutinosis* complex in the eastern United States. Illinois Biol. Mongr. **57**: 1-154.

Holcombe, T.L., Edgar, N.T. (1990): Late Cretaceous and Cenozoic evolution of Caribbean ridges and rises with special reference to paleogeography. In: Biogeographical Aspects of Insularity, p. 611-626. Azzaroli, A., Ed., Accademia Nazionale dei Lincei, Rome.

Iturralde-Vinent, M.A., MacPhee, R.D.E. (1999): Paleogeography of the Caribbean region: implications for Cenozoic biogeography. Bull. Amer. Mus. Nat. Hist. **238**: 1-95.

Iverson, J.B. (1978): Impact of feral cats and dogs on populations of West Indian rock iguana, *Cyclura carinata*. Biol. Conserv. **14**: 63-73.

Lawton, R.O., Nair, U.S., Pielke, R.A., Welch, R.M. (2001): Climatic impact of tropical lowland deforestation on nearby montane cloud forests. Science **294**: 584-587.

Lescure, J. (1979): Étude taxinomique et éco-éthologique d'un Amphibien des petites Antilles: *Leptodactylus fallax* Müller, 1926 (Leptodactylidae). Bull. Mus. Natn. Hist. Nat., Paris **1**: 757-774.

Malcolm, J.R., Liu, C.R., Neilson, R.P., Hansen, L., Hannah, L. (2006): Global warming and extinctions of endemic species from biodiversity hotspots. Cons. Biol. **20**: 538-548.

Matthew, W.D. (1915): Climate and evolution. Ann. NY Acad. Sci. **24**: 171-213.

Myers, N., Mittermeier, R.A., Mittermeier, C.G., da Fonseca, G.A.B., Kent, J. (2000): Biodiversity hotspots for conservation priorities. Nature **403**: 853-858.

Powell, R., Ottenwalder, J.A., Inchaustegui, S.J., Henderson, R.W., Glor, R.E. (2000): Amphibians and reptiles of the Dominican Republic: species of special concern. Oryx **34**: 118-128.

Raven, P.H. (2004): Taxonomy: Where are we now? Philos. T. Roy. Soc. B **359**: 729-730.

Reagan, D.P. (1992): Congeneric species distribution and abundance in a 3-dimensional habitat: the rain forest anoles of Puerto Rico. Copeia **1992**: 392-403.

Rivera, L.W., Aide, T.M. (1998): Forest recovery in the karst region of Puerto Rico. Forest Ecol. Manag. **108**: 63-75.

Rivera, L.W., Zimmerman, J.K., Aide, T.M. (2000): Forest recovery in abandoned agricultural lands in a karst region of the Dominican Republic. Plant Ecol. **148**: 115-125.

Rodda, G.H., Perry, G., Rondeau, R.J., Lazell, J. (2001): The densest terrestrial vertebrate. J. Trop. Ecol. **17**: 331-338.

Rosen, D.E. (1975): A vicariance model of Caribbean biogeography. Syst. Zool. **24**: 431-464.

Schuchert, C. (1935): Historical Geology of the Antillean-Caribbean region. John Wiley and Sons, New York.

Sergile, F.E., Woods, C.A. (2001): Status of conservation in Haiti: a 10-year retrospective. In: Biogeography of the West Indies: Patterns and Perspectives, p. 547-560. Woods, C.A., Sergile, F.E., Eds, CRC Press, Boca Raton, Florida.

Simpson, G.G. (1956): Zoogeography of West Indian land mammals. Amer. Mus. Novit. **1759**: 1-28.

Smith, M.L., Hedges, S.B., Buck, W., Hemphill, A., Inchaustegui, S., Ivie, M., Martina, D., Maunder, M., Ortega, J.F. (2005): Caribbean islands. In: Hotspots Revisited: Earth's Biologically Richest and Most Endangered Terrestrial Ecoregions. Mittermeier, R.A., Gil, P.R., Hoffmann, M., Pilgrim, J., Brooks, T., Mittermeier, C.G., Lamoreux, J., da Fonseca, G.A., Eds, CEMEX, Mexico City.

Stuart, S.N., Chanson, J.S., Cox, N.A., Young, B.E., Rodrigues, A.S.L., Fischman, D.L., Waller, R.W. (2004): Status and trends of amphibian declines and extinctions worldwide. Science **306**: 1783-1786.

Tole, L. (2002): Habitat loss and anthropogenic disturbance in Jamaica's Hellshire Hills area. Biodivers. Conserv. **11**: 575-598.

Tolson, P.J. (1996): Conservation of *Epicrates monensis* on the satellite islands of Puerto Rico. In: Contributions to West Indian Herpetology: a Tribute to Albert Schwartz, p. 407-416. Powell, R., Henderson, R.W., Eds, Society for the Study of Amphibians and Reptiles, Ithaca.

Trueb, L., Tyler, M.J. (1974): Systematics and evolution of the Greater Antillean hylid frogs. Occas. Pap. Mus. Nat. Hist. Univ. Kansas **24**: 1-60.

Vogel, P., Nelson, R., Kerr, R. (1996): Conservation strategy for the Jamaican iguana, *Cyclura collei*. In: Contributions to West Indian Herpetology: a Tribute to Albert Schwartz, p. 395-406. Powell, R., Henderson, R.W., Eds, Society for the Study of Amphibians and Reptiles, Ithaca.

Wallace, A.R. (1881): Island Life. Harper, New York.

Williams, E.E. (1989): Old problems and new opportunities in West Indian biogeography. In: Biogeography of the West Indies: Past, Present, and Future, p. 1-46. Woods, C.A., Ed., Sandhill Crane Press, Gainesville.

Wilson, E.O. (2004): Taxonomy as a fundamental discipline. Phil. Trans. Roy. Soc. B **359**: 739.

Woods, C.A., Sergile, F.E., Ottenwalder, J.A. (1992): Stewardship Plan for the National Parks and Nnatural Areas of Haiti. Florida Museum of Natural History, Gainesville, Florida.

WRI (2006): EarthTrends: The Environmental Information Portal. Available at http://earthtrends.wri.org. World Resources Institute, Washington, DC.

Young, B.E., Stuart, S.N., Chanson, J.S., Cox, N.A., Boucher, T.M. (2004): Disappearing Jewels: The Status of New World Amphibians. NatureServe, Arlington, Virginia.

Accepted: August 7, 2006 (BSW).
Reprinted from Applied Herpetology **3**: 281-292 (2006).

The conservation status of amphibians in the West Indies

S. Blair Hedges[1,2], Luis M. Díaz[3]

[1] Department of Biology, Pennsylvania State University, 208 Mueller Laboratory, University Park, PA 16802-5301, USA
[2] Corresponding author; e-mail: sbh1@psu.edu
[3] Museo Nacional de Historia Natural de Cuba, Apartado Postal 2349, La Habana 2, CP 10200, Cuba

Abstract. There are 196 species of amphibians known from the West Indies, 188 of which are native. With only a few exceptions, all of those native species are endemic to single islands or island banks and most are restricted to a small region within an island such as a single mountain top. The native species are members of the following families: Aromobatidae (1 species), Bufonidae (12 sp.), Hylidae (9 sp.), Eleutherodactylidae (161 sp.), Leptodactylidae (3 sp.), and Strabomantidae (2 sp.). The recent Global Amphibian Assessment found that 84% of West Indian amphibian species are threatened and that 71% are in the two highest threat categories, Endangered and Critically Endangered. The remote areas where many species occur constrain monitoring efforts, but remarkably, 49 species (26% of all native species) have not been observed in at least 10 years and 31 species (16%) in at least 20 years. Much-needed surveys will almost certainly find some of those species, but five species have not been found despite intensive search efforts, and are likely extinct — for unknown reasons. Small distributions and declining habitat area and quality were the most important considerations in determining conservation status. Many species are contained within national parks and other protected areas, but the effectiveness of the protected areas varies from essentially no protection to moderately good protection. Unfortunately, most of the critically endangered species occur in countries (e.g., Haiti) where protected areas have low effectiveness. Deforestation and habitat modification continue to be the most serious and widespread threats, although the impacts of an introduced fungal pathogen and climate change are being carefully monitored. For unexplained reasons, the largest conservation agencies, The Nature Conservancy and Conservation International, have yet to establish programs of any significance in countries of the Caribbean region (e.g., Haiti) where the biodiversity crisis is the most severe. Nearly all of the biological data used in determining the conservation status of species comes from the field of systematics and taxonomy rather than ecology, yet this fact has yet to be fully appreciated by the culture of conservation.

Key words: Antilles; biodiversity; Caribbean; deforestation; ecology; extinction; genetic resources; systematics; taxonomy.

Introduction

The West Indies are located between North and South America and are similar in total land area to Great Britain. There is some variability in how the biogeographic region is defined, but most biologists describe it as including the islands of the Bahamas Bank, the Greater Antilles (Cuba, Jamaica, Hispaniola, and Puerto Rico), the Lesser Antilles, and the Cayman Islands. It has become acceptable to refer to the same region as the "Caribbean Islands" (Smith et al., 2005), although technically that label is not correct because the islands of the Bahamas Bank are in the Atlantic Ocean, not the Caribbean Sea. Details of the physical geography of the West Indies and the historical and ecological biogeography of West Indian amphibians are reviewed elsewhere (Hedges, 1996, 1999, 2006b; Iturralde-Vinent and MacPhee, 1999).

The climate of the West Indies is influenced by prevailing winds from the northeast, bringing moisture primarily to the northern and eastern areas of each island. Forests originally covered virtually all parts of all islands, but approximately 90% of those original forests have been destroyed by humans (Hedges and Woods, 1993; Smith et al., 2005). Like most terrestrial organisms in the tropics, amphibians are essentially forest-dwelling animals. Therefore the decimation of the West Indian forests has greatly reduced the area of available habitats for amphibians, regardless of their abundance within existing habitats.

There are 196 species of amphibians known from the West Indies, 188 of which are native (Hedges, 2010; Henderson and Powell, 2009) (table 1). Additional undescribed species on Cuba, Jamaica, Hispaniola, and Puerto Rico are known to us, and we suspect that the actual number of West Indian amphibians, when all are finally discovered and described, is likely to exceed 300 species. All, or nearly all, of the native species are endemic to the region, and, with few exceptions, individual species are endemic to single islands or island banks. Many are restricted to a small area within an island or single mountain top. Of the native amphibians, there is one aromobatid (volcano frog), 12 bufonids (toads), 9 hylids (treefrogs), 161 eleutherodactylids (direct-developing terrestrial frogs), 3 leptodactylids (ditch frogs), and 2 strabomantids (direct-developing terrestrial frogs). Most species have been included in molecular phylogenies, and the taxonomy used here follows that in recent studies (Faivovich et al., 2005; Hedges and Heinicke, 2007; Hedges et al., 2008; Heinicke et al., 2007; Pramuk, 2006; Pramuk et al., 2001). Toad taxonomy continues to be debated; *Bufo* is used here but some prefer *Peltophryne* as a genus or subgenus.

The recent Global Amphibian Assessment (Stuart et al., 2004) concluded that 84% of West Indian amphibian species are threatened, being listed in the IUCN Redlist categories of Vulnerable, Endangered, and Critically Endangered (IUCN, 2010). This is the largest proportion of threatened species in any major amphibian fauna globally. The relatively small distributions of the species, small amount of remaining forests, and continuing threat from habitat destruction were major factors that led to the designation of such a large proportion of threatened species. Almost

Table 1. Native species of West Indian amphibians, their conservation status, and the most recent year that they were observed in the wild.

Species	Author of taxon	Status[1]	Last year observed[2]
Cuba			
Bufo cataulaciceps	Schwartz, 1959	EN	2009[3]
Bufo empusus	Cope, 1862	VU	2009[3-5,*]
Bufo florentinoi	Moreno and Rivalta, 2007	NA	2009[4-5,*]
Bufo fustiger	Schwartz, 1960	LC	2010[4-5]
Bufo gundlachi	Ruibal, 1959	VU	2009[3-5,*]
Bufo longinasus	Stejneger, 1905	EN	2009[3-5,*]
Bufo peltocephalus	Tschudi, 1838	LC	2010[4-5]
Bufo taladai	Schwartz, 1960	VU	2010[6]
Osteopilus septentrionalis	Duméril and Bibron, 1841	LC	2010[4-7]
Eleutherodactylus acmonis	Schwartz, 1960	EN	2010[6]
Eleutherodactylus adelus	Diaz, Cádiz, and Hedges, 2003	EN	2006[6]
Eleutherodactylus albipes	Barbour and Shreve, 1937	CR	2006[4-6]
Eleutherodactylus atkinsi	Dunn, 1925	LC	2010[6-7]
Eleutherodactylus auriculatus	Cope, 1862	LC	2010[6]
Eleutherodactylus bartonsmithi	Schwartz 1960	CR	2009[4-5]
Eleutherodactylus blairhedgesi	Estrada, Díaz, and Rodríguez, 1997	CR	2009[8]
Eleutherodactylus bresslerae	Schwartz, 1960	CR	2009[4-5]
Eleutherodactylus casparii	Dunn, 1926	EN	2009[4-5]
Eleutherodactylus cubanus	Barbour, 1942	CR	2007[4-6]
Eleutherodactylus cuneatus	Cope, 1862	LC	2010[6]
Eleutherodactylus dimidiatus	Cope, 1862	NT	2010[6]
Eleutherodactylus eileenae	Dunn, 1926	NT	2009[3-5]
Eleutherodactylus emiliae	Dunn, 1926	EN	2007[4-5]
Eleutherodactylus erythroproctus	Schwartz, 1960	NA	2007[4-5]
Eleutherodactylus etheridgei	Schwartz, 1958	EN	2009[6,*]
Eleutherodactylus glamyrus	Estrada and Hedges, 1997	EN	2007[4-6]
Eleutherodactylus goini	Schwartz, 1960	VU	2010[4-5]
Eleutherodactylus greyi	Dunn, 1926	EN	2009[4-5]
Eleutherodactylus guanahacabibes	Estrada and Novo Rodríguez, 1985	EN	2010[4-5]
Eleutherodactylus guantanamera	Hedges, Estrada, and Thomas, 1992	VU	2010[6]
Eleutherodactylus gundlachi	Schmidt, 1920	EN	2010[6]
Eleutherodactylus iberia	Estrada and Hedges, 1996	CR	2010[6]
Eleutherodactylus intermedius	Barbour and Shreve, 1937	EN	2009[4-5]
Eleutherodactylus ionthus	Schwartz, 1960	EN	2010[7]
Eleutherodactylus jaumei	Estrada and Alonso, 1997	CR	2007[4-5]
Eleutherodactylus klinikowskii	Schwartz, 1959	EN	2009[3-6,*]
Eleutherodactylus leberi	Schwartz, 1965	EN	2007[6]
Eleutherodactylus limbatus	Cope, 1862	VU	2010[4-5]
Eleutherodactylus maestrensis	Díaz, Cádiz, and Navarro, 2005	DD	2005[4-5,7]
Eleutherodactylus mariposa	Hedges, Estrada, and Thomas, 1992	CR	2007[3-6]
Eleutherodactylus melacara	Hedges, Estrada, and Thomas, 1992	EN	2010[7]
Eleutherodactylus michaelschmidi	Díaz, Cádiz, and Navarro, 2007	NA	2007[6]
Eleutherodactylus olibrus	Schwartz, 1958	NA	2010[4-5]

Table 1. (Continued.)

Species	Author of taxon	Status[1]	Last year observed[2]
Eleutherodactylus orientalis	Barbour and Shreve, 1937	CR	2010[6]
Eleutherodactylus pezopetrus	Schwartz, 1960	CR	2007[4-6]
Eleutherodactylus pinarensis	Dunn, 1926	EN	2009[3-5,*]
Eleutherodactylus planirostris	Cope, 1862	LC	2010[4-7]
Eleutherodactylus principalis	Estrada and Hedges, 1997	EN	2009[6]
Eleutherodactylus ricordii	Duméril and Bibron, 1841	VU	2009[4-6,*]
Eleutherodactylus riparius	Estrada and Hedges, 1998	LC	2010[6]
Eleutherodactylus rivularis	Díaz, Estrada, and Hedges, 2001	CR	2007[6]
Eleutherodactylus ronaldi	Schwartz, 1960	VU	2010[6]
Eleutherodactylus simulans	Díaz and Fong, 2001	EN	2010[6]
Eleutherodactylus staurometopon	Schwartz, 1960	NA	2008[6]
Eleutherodactylus symingtoni	Schwartz, 1957	CR	2010[6]
Eleutherodactylus tetajulia	Estrada and Hedges, 1996	CR	2010[6]
Eleutherodactylus thomasi	Schwartz, 1959	CR	2009[7,*]
Eleutherodactylus toa	Estrada and Hedges, 1991	EN	2010[6]
Eleutherodactylus tonyi	Estrada and Hedges, 1997	CR	2009[4-5,*]
Eleutherodactylus turquinensis	Barbour and Shreve, 1937	CR	2007[4-6]
Eleutherodactylus varians	Gundlach and Peters in Peters, 1864	VU	2009[3,*]
Eleutherodactylus varleyi	Dunn, 1925	LC	2010[4-5,7]
Eleutherodactylus zeus	Schwartz, 1958	EN	2009[3,6,*]
Eleutherodactylus zugi	Schwartz, 1958	EN	2009[6]
Jamaica			
Osteopilus brunneus	Gosse, 1851	LC	2010[9]
Osteopilus crucialis	Harlan, 1826	EN	2010[9]
Osteopilus marianae	Dunn, 1926	EN	2007[9]
Osteopilus wilderi	Dunn, 1925	EN	2009[10]
Eleutherodactylus alticola	Lynn, 1937	CR	1986[11]
Eleutherodactylus andrewsi	Lynn, 1937	EN	2006[12]
Eleutherodactylus cavernicola	Lynn, 1954	CR	2002[13]
Eleutherodactylus cundalli	Dunn, 1926	VU	2010[9]
Eleutherodactylus fuscus	Lynn and Dent, 1942	CR	1984[11]
Eleutherodactylus glaucoreius	Schwartz and Fowler, 1973	NT	2006[12]
Eleutherodactylus gossei	Dunn, 1926	LC	2009[10]
Eleutherodactylus grabhami	Dunn, 1926	EN	2007[12]
Eleutherodactylus griphus	Crombie, 1986	CR	1985[11]
Eleutherodactylus jamaicensis	Barbour, 1910	EN	1987[11]
Eleutherodactylus junori	Dunn, 1926	CR	1998[11]
Eleutherodactylus luteolus	Gosse, 1851	EN	2009[10]
Eleutherodactylus nubicola	Dunn, 1926	EN	1998[11]
Eleutherodactylus orcutti	Dunn, 1928	CR	1985[11]
Eleutherodactylus pantoni	Dunn, 1926	NT	2009[9-10]
Eleutherodactylus pentasyringos	Schwartz and Fowler, 1973	VU	2007[10-11]
Eleutherodactylus sisyphodemus	Crombie, 1977	CR	1984[11]

Table 1. (Continued.)

Species	Author of taxon	Status[1]	Last year observed[2]
Hispaniola			
Bufo fluviaticus	Schwartz, 1972	CR	1971[14]
Bufo fractus	Schwartz, 1972	EN	1969[14]
Bufo guentheri	Cochran, 1941	VU	2010[15-16]
Hypsiboas heilprini	Noble, 1923	VU	2010[15-16]
Osteopilus pulchrilineatus	Cope, 1869	EN	2010[15-16]
Osteopilus vastus	Cope, 1871	EN	2010[15-16]
Osteopilus dominicensis	Tschudi, 1838	LC	2009[11]
Eleutherodactylus abbotti	Cochran, 1923	LC	2010[15-16]
Eleutherodactylus alcoae	Schwartz, 1971	EN	2008[15-16]
Eleutherodactylus amadeus	Hedges, Thomas, and Franz, 1987	CR	1991[11]
Eleutherodactylus aporostegus	Schwartz, 1965	NA	2006[11,17,*]
Eleutherodactylus apostates	Schwartz, 1973	CR	2005[11,17,*]
Eleutherodactylus armstrongi	Noble and Hassler, 1933	EN	2007[18]
Eleutherodactylus audanti	Cochran, 1934	VU	2009[11,*]
Eleutherodactylus auriculatoides	Noble, 1923	EN	2009[15-16]
Eleutherodactylus bakeri	Cochran, 1935	CR	2006[11,17,*]
Eleutherodactylus bothroboans	Schwartz, 1965	NA	1986[11]
Eleutherodactylus brevirostris	Shreve, 1936	CR	2006[11,17,*]
Eleutherodactylus caribe	Hedges and Thomas, 1992	CR	1991[14,*]
Eleutherodactylus chlorophenax	Schwartz, 1976	CR	1985[11]
Eleutherodactylus corona	Hedges and Thomas, 1992	CR	1991[14,*]
Eleutherodactylus counouspeus	Schwartz, 1964	EN	1991[11]
Eleutherodactylus darlingtoni	Cochran, 1935	CR	1985[11]
Eleutherodactylus diplasius	Schwartz, 1973	NA	2006[11,17]
Eleutherodactylus dolomedes	Hedges and Thomas, 1992	CR	1991[14,*]
Eleutherodactylus eunaster	Schwartz, 1973	CR	2005[11,17,*]
Eleutherodactylus flavescens	Noble, 1923	NT	2010[15-16]
Eleutherodactylus fowleri	Schwartz, 1973	CR	1985[11]
Eleutherodactylus furcyensis	Shreve and Williams, 1963	CR	2009[11,*]
Eleutherodactylus glandulifer	Cochran, 1935	CR	1991[11,*]
Eleutherodactylus glanduliferoides	Shreve, 1936	CR	1985[11]
Eleutherodactylus glaphycompus	Schwartz, 1973	EN	2006[11,17,*]
Eleutherodactylus grahami	Schwartz, 1979	EN	1991[11]
Eleutherodactylus haitianus	Barbour, 1942	EN	2009[15-16]
Eleutherodactylus heminota	Shreve and Williams, 1963	EN	2009[15-16,*]
Eleutherodactylus hypostenor	Schwartz, 1965	EN	1984[11]
Eleutherodactylus inoptatus	Barbour, 1914	LC	2010[15-16]
Eleutherodactylus jugans	Cochran, 1937	CR	2009[11]
Eleutherodactylus lamprotes	Schwartz, 1973	CR	2005[11,17]
Eleutherodactylus leoncei	Shreve and Williams, 1963	CR	2007[18]
Eleutherodactylus limbensis	Lynn, 1958	NA	1953[14]
Eleutherodactylus lucioi	Schwartz, 1980	CR	1979[14]
Eleutherodactylus melatrigonum	Schwartz, 1966	NA	1993[11]
Eleutherodactylus minutus	Noble, 1923	EN	2009[15-16]

Table 1. (Continued.)

Species	Author of taxon	Status[1]	Last year observed[2]
Eleutherodactylus montanus	Schmidt, 1919	EN	2007[18]
Eleutherodactylus nortoni	Schwartz, 1976	CR	2006[11,17,*]
Eleutherodactylus notidodes	Schwartz, 1966	NA	1999[11]
Eleutherodactylus oxyrhynchus	Duméril and Bibron, 1841	CR	1991[11]
Eleutherodactylus parabates	Schwartz, 1964	CR	1996[11]
Eleutherodactylus paralius	Schwartz, 1976	NA	1969[14]
Eleutherodactylus parapelates	Hedges and Thomas, 1987	CR	1984[14,*]
Eleutherodactylus patriciae	Schwartz, 1964 (1965)	EN	2009[15-16]
Eleutherodactylus paulsoni	Schwartz, 1964	CR	1991[11]
Eleutherodactylus pictissimus	Cochran, 1935	VU	2009[15-16,*]
Eleutherodactylus pituinus	Schwartz, 1964	EN	2009[15-16]
Eleutherodactylus poolei	Cochran, 1938	CR	1985[11]
Eleutherodactylus probolaeus	Schwartz, 1965	EN	1985[11]
Eleutherodactylus rhodesi	Schwartz, 1980	CR	1985[11]
Eleutherodactylus rucillensis	Cochran, 1939	NA	1986[11]
Eleutherodactylus rufifemoralis	Noble and Hassler, 1933	CR	1999[15-16]
Eleutherodactylus ruthae	Noble, 1923	EN	1963[14]
Eleutherodactylus schmidti	Noble, 1923	CR	1986[11]
Eleutherodactylus sciagraphus	Schwartz, 1973	CR	1984[11]
Eleutherodactylus semipalmatus	Shreve, 1936	CR	1985[19]
Eleutherodactylus sommeri	Schwartz, 1977	NA	2009[15-16]
Eleutherodactylus thorectes	Hedges, 1988	CR	1991[11,*]
Eleutherodactylus tychathrous	Schwartz, 1965	NA	1986[11]
Eleutherodactylus ventrilineatus	Shreve, 1936	CR	2006[11,17,*]
Eleutherodactylus warreni	Schwartz, 1976	CR	1970[14]
Eleutherodactylus weinlandi	Barbour, 1914	LC	2010[15-16]
Eleutherodactylus wetmorei	Cochran, 1932	VU	2009[11,*]
Leptodactylus albilabris	Günther, 1859	LC	2009[15-16]
Bahamas			
Osteopilus septentrionalis	Duméril and Bibron, 1841	LC	1993[11]
Eleutherodactylus planirostris	Cope, 1862	LC	1993[11]
Eleutherodactylus rogersi	Goin, 1955	NA	1993[11]
Puerto Rico Region			
Bufo lemur	Cope, 1868	CR	2009[20]
Eleutherodactylus antillensis	Reinhardt and Lütken, 1862	LC	2010[20-22]
Eleutherodactylus brittoni	Schmidt, 1920	LC	2010[20-21]
Eleutherodactylus cochranae	Grant, 1932	LC	2010[20-22]
Eleutherodactylus cooki	Grant, 1931	EN	2009[20]
Eleutherodactylus coqui	Thomas, 1966	LC	2010[20-22]
Eleutherodactylus eneidae	Rivero, 1959	CR	1990[11]
Eleutherodactylus gryllus	Schmidt, 1920	EN	2010[20-21]
Eleutherodactylus hedricki	Rivero, 1963	EN	2010[20-21]
Eleutherodactylus jasperi	Drewry and Jones, 1976	CR	1981[23]

Table 1. (Continued.)

Species	Author of taxon	Status[1]	Last year observed[2]
Eleutherodactylus juanariveroi	Rios-López and Thomas, 2007	NA	2010[20]
Eleutherodactylus karlschmidti	Grant, 1931	CR	1988[20]
Eleutherodactylus lentus	Cope, 1862	EN	2010[22]
Eleutherodactylus locustus	Schmidt, 1920	CR	2010[20-21]
Eleutherodactylus monensis	Meerwarth, 1901	VU	2006[21-24]
Eleutherodactylus portoricensis	Schmidt, 1927	EN	2010[20-21]
Eleutherodactylus richmondi	Stejneger, 1904	CR	2010[20-21]
Eleutherodactylus schwartzi	Thomas, 1966	EN	2010[20]
Eleutherodactylus unicolor	Stejneger, 1904	CR	2010[20-21]
Eleutherodactylus wightmanae	Schmidt, 1920	EN	2010[20-21]
Leptodactylus albilabris	Günther, 1859	LC	2010[20-22]
Lesser Antilles			
Allobates chalcopis	Kaiser, Coloma, and Gray, 1994	VU	1992[14]
Eleutherodactylus amplinympha	Kaiser, Green, and Schmid, 1994 (1995)	EN	2008[25]
Eleutherodactylus barlagnei	Lynch, 1965	EN	1993[26]
Eleutherodactylus johnstonei	Barbour, 1914	LC	2010[27]
Eleutherodactylus martinicensis	Tschudi, 1838	NT	2009[11]
Eleutherodactylus pinchoni	Schwartz, 1967	EN	2001[26]
Pristimantis euphronides	Schwartz, 1967	EN	2010[27]
Pristimantis shrevei	Schwartz, 1967	EN	2006[25,28]
Leptodactylus fallax	Muller, 1926	CR	2007[29]
Leptodactylus validus	Garman, 1887	LC	2007[29]
Cayman Islands			
Osteopilus septentrionalis	Duméril and Bibron, 1841	LC	2010[30]
Eleutherodactylus planirostris	Cope, 1862	LC	2010[30]

[1]IUCN Redlist categories: CR (Critically Endangered), DD (Data Deficient), EN (Endangered), LC (Least Concern), NA (Not Assessed), NT (Near Threatened), and VU (Vulnerable). [2]Observations include collection, visual, or vocalization records, and include data and personal communications from: [3]Roberto Alonso, [4]Luis M. Díaz, pers. obs., [5]Antonio Cádiz, [6]Ariel Rodriguez, [7]Ansel Fong, [8]Javier Torres, [9]Susan Koenig, [10]Alejandro Sánchez, [11]S. Blair Hedges, pers. obs., [12]Miguel Nelson, [13]Byron Wilson, [14]Original description or subsequent literature, [15]Sixto Inchaustegui, [16]Marcelino Hernández, [17]Eladio Fernández, [18]Marcos Rodríguez, [19]University of Florida, Museum of Natural History expedition, [20]Alejandro Sánchez, [21]Patricia Burrowes, [22]Renata Platenberg, [23]Richard Thomas, [24]Rafael Joglar, [25]Robert Powell, [26]Breuil (2002), [27]Craig Berg, [28]Robert W. Henderson, [29]Jay D. King, [30]Fred Burton. *See note added in proof.

as confirmation of the process, a recent statistical analysis of factors associated with IUCN threat status, among all amphibians, concluded that species with small geographic ranges were mostly likely to have high threat risk (Sodhi et al., 2008). However, it is the additional factor of declining habitat — even within many so-called protected areas — that is responsible for such a high proportion of threatened species in the West Indies.

The conservation of West Indian amphibians has been addressed in a large number of recent reviews. Most of these have appeared in the journal *Applied Herpetology*, including general reviews (Hedges, 2006a; Wilson et al., 2006) and others focused on specific islands or groups of islands: Antigua, Barbuda, and Redonda (Daltry, 2007), Dominica (Malhotra et al., 2007), the Dutch Windward Islands (Powell, 2006), the French West Indies (Lorvelec et al., 2007), Grenada and the Grenadines (Daudin and de Silva, 2007; Henderson and Berg, 2006), Puerto Rico (Joglar et al., 2007), and the Virgin Islands (Perry and Gerber, 2006; Platenberg and Boulon, 2006). Besides those reviews, one of us (S.B.H.) also has reviewed the West Indian species of amphibians elsewhere (Hedges, 2008), and, with other authors, the Neotropical species of amphibians in general (Bolaños et. al., 2008). In addition, S.B.H. has reviewed conservation issues related to the direct-developing frogs of the New World (Hedges et al., 2008), which comprise 86% of the West Indian amphibian fauna.

This unusual attention has made the West Indies one of the best known regions of the world in terms of knowledge of the conservation status of its amphibian fauna. This does not mean that all of the data needed are in hand, only that the available information has been summarized and the gaps in our knowledge have been identified. For this reason, we will not attempt to duplicate what has been reviewed elsewhere. Instead, we provide a brief overview here of the conservation status of the species, threats to their survival, and recommendations. Most importantly, we include a species census of the West Indian amphibian fauna listing the most recent year that a species has been observed in the wild. This provides an update of similar lists published by one of us (S.B.H.) in previous years (Hedges, 1993, 1999). Such a list can help focus field efforts in tracking species of concern.

Conservation Status of Species

The recent review of the conservation status of amphibians (Stuart et al., 2004; IUCN, 2010) was done in a systematic fashion using established criteria, and was reviewed by herpetologists who conduct research in the West Indies and are familiar with particular species and groups of species. The major review was made in 2004 and some species accounts have since been updated. It is the best available review of the status of species globally. In it, 84% of the total native species in the region (169 species at the time of assessment) were found to be threatened. Of those, 64 were listed as Critically Endangered, 60 as Endangered, and 18 as Vulnerable (table 1). Another six species were considered to be Near Threatened. There is no other region in the world with such a high proportion of threatened species. It is especially worrisome that more than one-third (37%) of the native species of West Indian amphibians were placed in the highest threat category, Critically Endangered. To determine that a species of amphibian, especially a small frog, is actually extinct is difficult because of the complexity and remoteness of the habitat and near-impossibility of examining every possible place where an individual might exist. For this reason, the IUCN designation of "possibly extinct" is appropriate in these cases of critically endangered species that have not been seen in many years and may be extinct.

Since that assessment in 2004, 17 additional new species, including those elevated from subspecies, were added to the list of amphibian species native to the West Indies (table 1). None has yet to be assessed for conservation status under IUCN guidelines. Moreover, any species which has been affected by the elevation of a subspecies must be reassessed because — by definition — the distribution of that species is now smaller.

In addition to listing the conservation status of each native species in table 1, we note the year of the most recent observation of a species in the wild. Unlike highly visible animals such as birds, tropical amphibians tend to be cryptic and it is not unusual for a species to go unnoticed for years or decades, simply because no one has looked for it. Our table reveals this problem, especially in the case of the Bahamas, Hispaniola, and Jamaica. The original species census for the West Indies in 1993 (Hedges, 1993), updated in 1999 (Hedges, 1999) was fairly up-to-date at that time, because the senior author (S.B.H.) and colleagues had conducted extensive field work on many islands during the 1980s and early 1990s. However, field work on Jamaica and Hispaniola by herpetologists has focused more on individual species rather than faunal surveys in the last 15 years. Consequently, some species on those two large islands have not been sought in two decades. Fortunately, recent field work by the junior author (L.M.D.) and colleagues in Cuba has provided an up-to-date census for the species of that island. Likewise, the resident amphibian biologists of Puerto Rico (e.g., Patricia Burrowes, Raphael Joglar, Alejandro Sánchez, and Richard Thomas), have given us the latest information for that island.

While acknowledging biases in these census data, there are nonetheless 49 species that have not been seen in more than 10 years: 8 on Jamaica, 36 on Hispaniola, 3 on Puerto Rico, 1 in the Bahamas (2 others not seen also occur on Cuba so are not counted here), and 1 in the Lesser Antilles. Even more significantly, 31 species have not been seen in more than 20 years: 6 on Jamaica, 22 on Hispaniola, and 3 on Puerto Rico. Of these species there are at least five that apparently have disappeared, because searches have failed to locate them. Three occur on Puerto Rico (*Eleutherodactylus eneidae, E. jasperi*, and *E. karlschmidti*), one (*E. semipalmatus)* occurs on Hispaniola, and one (*E. orcutti*) occurs on Jamaica. Of the remaining species that have not been seen in 10 or 20 years, there may be some that have disappeared as well, but there are insufficient data from searches and surveys to draw any definite conclusions.

Of the five species that are possibly extinct, three (*Eleutherodactylus karlschmidti, E. semipalmatus*, and *E. orcutti*) occur in and around streams. A fourth species, *E. jasperi*, is an obligate bromeliad-dweller, and the fifth species, *E. eneidae*, is poorly known but is arboreal and occurs in cloud and elfin forest. This high proportion of riparian species (only 7% of all native West Indian species are riparian) is consistent with other regions, especially Middle America, where such species show the most declines (Campbell, 1999; McCranie and Wilson, 2002; Savage, 2002; Bolaños et al., 2008).

One of us (S.B.H) has searched extensively — without success — for all five "missing" species in habitats and localities where they once occurred in abundance. In addition, others have searched for *E. karlschmidti* extensively (Joglar et al., 2007), also without success. The current status of the riparian species of Guadeloupe, *E. barlagnei*, is unknown, but concern was expressed that populations were declining in the 1990s (Breuil, 2002). Curiously, other riparian species on Hispaniola (*Hypsiboas heilprini*, *Osteopilus vastus*, *E. schmidti*) and Cuba (*E. cuneatus, E. riparius, E. rivularis*, and *E. turquinensis*) have not disappeared, although none of these species has been monitored at the level of populations and therefore declines in abundance may be taking place.

In November, 2009, S.B.H. conducted a brief survey of amphibians in the Massif de la Selle of Haiti. Some amphibians were found but two critically endangered species, *Eleutherodactylus darlingtoni* and *E. gladuliferoides*, were not encountered at localities where they were found in previous years. Also, in December, 2009, he searched for *Allobates chalcopis* at known localities on Martinique but was unable to find that species. Conditions were somewhat dry on both islands, but it raises some concern and interest in additional, more extensive, surveys.

To summarize the conservation status of the species, most (84%) of the assessed species are threatened and more than two thirds (71%) of the assessed species are either Endangered or Critically Endangered; five of those are possibly extinct. We are concerned that other species that have not been observed in recent years, and which have limited ranges, also may have disappeared.

Threats to Survival

Habitat destruction

Forests originally covered all, or nearly all, of the West Indies before humans arrived. Therefore, as might be expected, amphibians are adapted to forest habitats. Even when a canopy is maintained but only the number of tree species is reduced (e.g., in coffee plantations) the diversity of amphibian species is considerably reduced (CNAP, 2003; Fong, 1999). Forests comprised of only a single species (e.g., of *Pinus* for lumber production), which are sometimes mistakenly counted in tabulations of forest cover, are especially poor habitats for amphibians.

Published values of forest cover for countries often vary because of different ways that the data are collected and analyzed. For example, an area may be scored as forested if as little as 50% or as much as 90% of a cell contains tree canopies. Obviously, if a value of 90% is used, a smaller total area would be scored as forested than if a value of 50% were used. Also, spectral data from satellite imagery may not be interpreted correctly as, for example, forest versus plantation, or primary (original) versus secondary forest. Islands in the West Indies often have many introduced tree species (mango, breadfruit, bamboo, etc.) and these frequently give the impression to a casual observer (and sometimes to compilers of forest cover data), erroneously, that the area is well-forested. Despite this sometimes large margin of error, most surveys have agreed that the West Indies has lost about 90% of its original forests (Hedges, 2006a; Hedges and Woods, 1993; Smith et al., 2005). One small island in the Lesser Antilles, Dominica, has a large amount of original forest but most of the land area, and amphibian species, in the West Indies are in the Greater Antilles and this is where forests have suffered the most. For example, it is doubtful whether Haiti has any remaining primary forest other than in isolated small patches. Entire mountains are essentially denuded, leaving almost no habitat for any organism aside from the decomposers.

Assuming the species-area relationship (MacArthur and Wilson, 1967), a 90% reduction in habitat will lead to a 50% reduction in the number of species. Table 1 indicates that, fortunately, such a prediction has not yet been realized. Why? First, that mathematical prediction does not take into account any other variables, such as the location of the remaining 10% of habitat; in this case it is mostly in upland areas where diversity is highest. Also, it is an equilibrium value, and it is unknown how much time is required to reach equilibrium. The concept of a “lag time” is a realistic assumption and can be seen in Haiti today. In that country, small, isolated patches of forest remain, which barely contribute to any total amount of forest cover. Yet sometimes in those small patches species may persist, temporarily, until even the patches are destroyed. Even if the isolated patches are not destroyed, the quantity or quality of the habitat may be insufficient for many species to persist. Species like this that still exist, but are unlikely to persist, have been referred to as the “living dead” (Wilson, 1992).

The causes of habitat destruction are complex and vary depending on the island, country, population size, and economy. Puerto Rico and Barbados are among the most densely populated areas on Earth. Because of their relatively strong economies, a major pressure on habitat is from urbanization (e.g., roads, buildings, houses, and parking lots). In contrast, destruction of forests in Haiti has been fueled largely by basic subsistence needs, especially energy (charcoal) for cooking food. Initially forests in the West Indies were cut for building materials and agriculture, especially sugar cane, and thus the reasons for habitat destruction have changed with time and levels of forest cover.

Threats other than habitat destruction

Other major threats to West Indian amphibians include a fungal pathogen, introduced predators, and climate change. The fungal pathogen is a chytrid, *Batrachochytrium dendrobatidis*, and it has been a focus of attention for global amphibian declines for the last decade (Collins and Crump, 2009). Recent work has identified how the fungus kills amphibians: by disruption of cutaneous function, including transport of electrolytes (Voyles et al., 2009).

The source region for the fungus is not known with certainty but it is believed to have been introduced to the New World by humans. Mass mortality of amphibians at localities in Panama following the appearance of the fungus has suggested that it is the proximal cause of declines (Lips et al., 2006). However, it is known to coexist with other amphibian faunas without causing declines (Daszak et al., 2005). Recently, two studies have implicated climate change as a factor in declines. In one, it was suggested that warmer climate favored the spread of the disease (Pounds et al., 2006) while in the other it was suggested that climate change alone, without help from the fungus, was responsible for declines through reduction in leaf litter (Whitfield et al., 2007). The relative importance of the chytrid pathogen, climate change, and habitat destruction in amphibian declines continues to be debated (Collins and Crump, 2009).

In the West Indies, the chytrid fungus has been reported in Puerto Rico (Burrowes et al., 2004), Hispaniola (Joglar et al., 2007), Cuba (Díaz et al., 2007), Dominica (Malhotra et al., 2007), Grenada (Henderson and Berg, 2006), and Montserrat (Durrell Wildlife Conservation Trust, unpublished reports). In the broader region of Caribbean islands it has turned up on Tobago as well (Alemu et al., 2008). In Puerto Rico, it seems to be restricted to upland areas and has not yet been found in lowland frog populations (Joglar et al., 2007). The three species of *Eleutherodactylus* in Puerto Rico that have disappeared could not be more different in ecological habits (streamside, bromeliad-dwelling, and arboreal), which makes it difficult to see an obvious link between the fungus and their declines. In a two-year study of *Eleutherodactylus coqui* and *E. portoricensis*, Longo et al. (2009) found that populations were persisting with the chytrid, but drier periods would lead to higher infection levels. In Cuba, the chytrid was discovered in October of 2006 in a dying individual of the endemic toad *Bufo longinasus dunni*. Although there is no current evidence of amphibian declines in Cuba (table 1), populations should be monitored carefully in the future. In Dominica, the chytrid is believed to be responsible for declines in the aquatic-breeding *Leptodactylus fallax*, although other species of frogs (non-aquatic) on the island have apparently not declined (Malhotra et al., 2007).

Over the centuries, humans — including indigenous peoples — have introduced various species of organisms to areas where they were not native. In the West Indies, the Black Rat (*Rattus rattus*) and Norway Rat (*R. norvegicus*) probably first arrived with Columbus in the late 15th Century, followed soon after by cats (*Felis catus*) and dogs (*Canis familiaris*). The Small Indian Mongoose (*Urva auropunctata*) was introduced in the mid-19th Century. All of these mammals, and others, are known to include amphibians in their diet. It is widely believed that at least some species of the native West Indian herpetofauna have become extinct because of such introduced predators (Henderson, 1992).

Because of the broad distribution of these and other introduced predators, there is no place in the West Indies that can be considered "virgin" habitat. For example, *Rattus rattus* is abundant in the uncut elfin forest on El Yunque (Puerto Rico) and is frequently seen climbing around in the canopy at night searching for food, potentially within bromeliads far above the ground. They are so abundant that on one evening an individual rat tripped on a branch, falling from the canopy to the head of the senior author who was recording anuran vocalizations! Also, frogs (*Eleutherodactylus* sp.) have been found in the stomach contents of mongooses from a forested area in eastern Puerto Rico (Viella, 1998). It is not known when these predators successfully colonized the upland forests of Puerto Rico, from the cities and lowlands, but the possibility exists that they may have been a factor in the decline of some species of frogs on that island.

In Cuba, the introduced African catfish (*Clarias gariepinus*) is a potential predator of eggs and tadpoles of those species that breed in water (*Bufo* and *Osteopilus*). The ecological plasticity, reproductive rate, and voracity of these fish, that can survive in a diversity of brackish and freshwater environments (including ditches and streams), make it a serious threat to the native biota. Since 2002, the African catfish has covered the entire island is now present in most rivers and in subterranean waters. Also in Cuba, pigs are a problem for at least one population of toad, *Bufo longinasus longinasus*, at Alturas de Pizarras del Sur, Pinar del Río province (Díaz and Cádiz, 2008), because they are continuously stirring up the margins of the small streams where the toads breed, killing metamorphs, disturbing adults, and clouding the water.

Conclusions and Recommendations

It is easy to review, as we have done here, what is known about the status of West Indian amphibians. But what is needed now are actual data on population levels (Joglar et al., 2007), potential threats (Burrowes et al., 2004; Díaz et al., 2007), and systematic surveys that provide information on the status and ecological health of species or entire faunas. The relationship of the chytrid fungus to amphibian declines continues to be a mystery that needs to be solved. Regardless of the outcome, most species of West Indian amphibians will continue to decline as their habitats decline.

In the West Indies, humans have modified amphibian habitats more than in any other part of the tropics, except for perhaps Madagascar and a few other countries that have similarly low levels of original forest. As the forests disappear, so go the amphibians and other forest-dwelling organisms. This is a fact, and this habitat destruction continues even within most parks and reserves in the Greater Antilles (Puerto Rico being a notable exception). In most cases, laws exist to prevent deforestation but they are often not enforced because of lack of resources.

Aid provided by the governments of developed countries usually is not distributed to developing countries in an innovative or effective manner that would help slow or halt declines of species. Instead, it is the large, non-profit, private organizations, such as The Nature Conservancy and Conservation International, which are in an ideal position to target conservation hot-spots and achieve rapid results. However, despite their active involvement and success in general conservation in the West Indies, both organizations have been slow to respond to the looming extinction crisis in the region.

For example, the most serious crisis for the biodiversity of the West Indies, and one of the most serious of any region in the world, is Haiti. Essentially no original forest remains and the existing biodiversity is literally hanging on by a thread, in small remaining and isolated patches of forest. Dozens of species of amphibians that occur nowhere else, not even in the neighboring Dominican Republic, are threatened with extinction. Haiti has established national parks and they have laws that protect forests and wildlife, but demands on resources within the parks are too great. A modest amount of assistance from private organizations such as The Nature Conservancy and Conservation International, for salaries of park guards and some infrastructure, could make a significant difference in addressing this crisis. The cutting of trees in parks, which is the immediate problem, needs to be stopped immediately. Also, to prepare for the worst, species need to be retrieved for frozen cell cultures (to be used in future whole-animal cloning) and captive breeding before they disappear entirely. For such crises, it is acceptable to focus on the short-term and proximal problems, even if there is no clear long-term solution, because it buys time for species survival.

Also, the conservation value of systematic (biotic) surveys is underappreciated. Systematists are the conservationists of the 21st Century. The great majority of information on the conservation status of amphibians reviewed in the Global Amphibian Assessment by the IUCN (Stuart et al., 2004; IUCN, 2010) was not funded by ecology or conservation grants. It was largely based on work conducted by systematists. Likewise this is true for nearly all West Indian species (table 1): their listing as Endangered, Critically Endangered, etc. is almost entirely based on systematic data (museum specimens and other data from systematists). For years, E.O. Wilson and others have made this same point (Wilson, 1992, 2000, 2004), about the under-appreciation of systematics for conservation, but private conservation agencies in general have been reluctant to fund such systematic surveys. Conservation International's Rapid Assessment Program is a notable exception. More of these programs are needed, and an obvious target in the West Indies would be Haiti.

Acknowledgements. We thank those colleagues who have shared their observations with us on the status of West Indian amphibians and reptiles, including Roberto Alonso, Craig Berg, Patricia Burrowes, Fred Burton, Antonio Cádiz, Eladio Fernández, Ansel Fong, Robert Henderson, Marcelino Hernández, Sixto Inchaustegui, Jay King, Gad Perry, Rafael Joglar, Susan Koenig, Miguel Landestoy, Miguel Nelson, Robert Ortiz, Renata Platenberg, Robert Powell, Neftalí Rios-López, Rayner Nuñez, Ariel Rodríguez, Marcos Rodríguez, Alejandro Sánchez, Richard Thomas, and Byron Wilson. The senior author thanks the National Science Foundation for support; the junior author thanks WWF (Canada) and the Linnean Society (UK).

References

Alemu, J.B., Cazabon, M.N.E., Dempewolf, L., Hailey, A., Lehtinen, R.M., Mannette, R.P., Naranjit, K.T., Roach, A.C.J. (2008): Presence of the chytrid fungus *Batrachochytrium dendrobatidis* in populations of the critically endangered frog *Mannophryne olmonae* in Tobago, West Indies. Ecohealth **5**: 34-39.

Bolaños, F., Castro, F., Cortez, C., De la Riva, I., Grant, T., Hedges, S.B., Heyer, W.R., Ibáñez, R., La Marca, E., Lavilla, E., Silvano, D.L., Lötters, S., Parra Olea, G., Reichle, S., Reynolds, R., Rodríguez, L., Santos Barrera, G., Scott, N.J., Ubeda, C., Veloso, A., Wilkinson, M., Young, B. (2008): Amphibians of the Neotropical Realm. In: Threatened Amphibians of the World, p. 92-99. Stuart, S.N., Hoffmann, M., Chanson, J.S., Cox, N.A., Berridge, R., Ramani, P., Young B.E., Eds, Barcelona, Lynx Ediciones.

Breuil, M. (2002): Histoire Naturelle des Amphibiens et Reptiles Terrestres de l'archipel Guadeloupéen. Paris, Muséum National d'Histoire Naturelle.

Burrowes, P.A., Joglar, R.L., Green, D.E. (2004): Potential causes for amphibian declines in Puerto Rico. Herpetologica **60**: 141-154.

Campbell, J.A. (1999): Distribution patterns of amphibians in Middle America. In: Patterns of Distribution of Amphibians: A Global Perspective, p. 111-210. Duellman, W.E., Ed., Baltimore, Maryland, Johns Hopkins Press.

CNAP (2003): Sistema Nacional de Áreas Protegidas de Cuba, Plan 2003-2008. Sevilla, Spain, Scandon Impresores.

Collins, J.P., Crump, M.L. (2009): Extinction in Our Times. New York, Oxford University Press.

Daltry, J.C. (2007): An introduction to the herpetofauna of Antigua, Barbuda and Redonda, with some conservation recommendations. Appl. Herpetol. **4**: 97-130.

Daszak, P., Scott, D.E., Kilpatrick, A.M., Faggioni, C., Gibbons, J.W., Porter, D. (2005): Amphibian population declines at savannah river site are linked to climate, not chytridiomycosis. Ecology **86**: 3232-3237.

Daudin, J., de Silva, M. (2007): An annotated checklist of the amphibians and terrestrial reptiles of the Grenadines with notes on their local natural history and conservation. Appl. Herpetol. **4**: 163-175.

Díaz, L.M., Cádiz, A. (2008): Guía taxonómica de los anfibios de Cuba. Abc Taxa **4**: 1-294 (+CD).

Díaz, L.M., Cádiz, A., Chong, A., Silva, A. (2007): First report of chytridiomycosis in a dying toad (Anura: Bufonidae) from Cuba: A new conservation challenge for the island. EcoHealth **4**: 172-175.

Faivovich, J., Haddad, C.F.B., Garcia, P.C.A., Frost, D.R., Campbell, J.A., Wheeler, W.C. (2005): Systematic review of the frog family hylidae, with special reference to hylinae: phylogenetic analysis and taxonomic revision. Bull. Am. Mus. Nat. Hist. **294**: 1-240.

Fong, A. (1999): Changes in amphibian composition in altered habitats of eastern Cuba. Froglog **36**: 2-3.

Hedges, S.B. (1993): Global amphibian declines: A perspective from the Caribbean. Biodivers. Conserv. **2**: 290-303.

Hedges, S.B. (1996): The origin of West Indian amphibians and reptiles. In: Contributions to West Indian Herpetology: A Tribute to Albert Schwartz, p. 95-127. Powell, R., Henderson, R.W., Eds, Ithaca, New York, Society for the Study of Amphibians and Reptiles.

Hedges, S.B. (1999): Distribution patterns of amphibians in the West Indies. In: Patterns of Distribution of Amphibians: A Global Perspective, p. 211-254. Duellman, W.E., Ed., Baltimore, Maryland, Johns Hopkins Press.

Hedges, S.B. (2006a): An overview of the evolution and conservation of West Indian amphibians and reptiles. Appl. Herpetol. **3**: 281-292.

Hedges, S.B. (2006b): Paleogeography of the Antilles and origin of West Indian terrestrial vertebrates. Ann. Mo. Bot. Gard. **93**: 231-244.

Hedges, S.B. (2008): Threatened amphibians of the West Indies. In: Threatened Amphibians of the World, p. 104-105. Stuart, S.N., Hoffmann, M., Chanson, J.S., Cox, N.A., Berridge, R., Ramani, P., Young B.E., Eds, Barcelona, Lynx Ediciones.

Hedges, S.B. (2010): Caribherp: Database of West Indian amphibians and reptiles (http://www.caribherp.org/, accessed on 20 April 2010). Pennsylvania State University, University Park, Pennsylvania.

Hedges, S.B., Heinicke, M.P. (2007): Molecular phylogeny and biogeography of West Indian frogs of the genus *Leptodactylus*. Mol. Phylogenet. Evol. **44**: 308-314.

Hedges, S.B., Duellman, W.E., Heinicke, M.P. (2008): New World direct-developing frogs (Anura: Terrarana): Molecular phylogeny, classification, biogeography, and conservation. Zootaxa **1737**: 1-182.

Hedges, S.B., Woods, C.A. (1993): Caribbean hot spot. Nature **364**: 375.

Heinicke, M.P., Duellman, W.E., Hedges, S.B. (2007): Major Caribbean and Central American frog faunas originated by oceanic dispersal. Proc. Natl. Acad. Sci. USA **104**: 10092-10097.

Henderson, R.W. (1992): Consequences of predator introductions and habitat destruction on amphibians and reptiles in the post-Columbus West Indies. Carib. J. Sci. **28**: 1-10.

Henderson, R.W., Powell, R. (2009): Natural History of West Indian Reptiles and Amphibians. Gainesville, Florida, University Press of Florida.

Henderson, R.W., Berg, C.S. (2006): The herpetofauna of Grenada and the Grenada Grenadines: Conservation concerns. Appl. Herpetol. **3**: 197-213 [and 2010 addendum in this volume].

Iturralde-Vinent, M.A., MacPhee, R.D.E. (1999): Paleogeography of the Caribbean region: Implications for Cenozoic biogeography. Bull. Am. Mus. Nat. Hist. **238**: 1-95.

IUCN (2010): IUCN Redlist of Threatened Species (http://www.iucnredlist.org/initiatives/amphibians, accessed on 19 April 2010). Cambridge, United Kingdom International Union for the Conservation of Nature.

Joglar, R.L., Álvarez, A.O., Aide, T.M., Barber, D., Burrowes, P.A., García, M.A., León-Cardona, A., Longo, A.V., Pérez-Buitrago, N., Puente, A., Rios-López, N., Tolson, P.J. (2007): Conserving the Puerto Rican herpetofauna. Appl. Herpetol. **4**: 327-345.

Lips, K.R., Brem, F., Brenes, R., Reeve, J.D., Alford, R.A., Voyles, J., Carey, C., Livo, L., Pessier, A.P., Collins, J.P. (2006): Emerging infectious disease and the loss of biodiversity in a Neotropical amphibian community. Proc. Natl. Acad. Sci. USA **103**: 3165-3170.

Longo, A.V., Burrowes, P.A., Joglar, R.L. (2009): Seasonality of *Batrachochytrium dendrobatidis* infection in direct-developing frogs suggests a mechanism for persistence. Dis. Aquat. Org. (published online; doi: 10.3354/dao02054).

Lorvelec, O., Pascal, M., Pavis, C., Feldmann, P. (2007): Amphibians and reptiles of the French West Indies: Inventory, threats and conservation. Appl. Herpetol. **4**: 131-161.

MacArthur, R.H., Wilson, E.O. (1967): The theory of island biogeography. Princeton, New Jersey, Princeton University Press.

Malhotra, A., Thorpe, R.S., Hypolite, E., James, A. (2007): A report on the status of the herpetofauna of the Commonwealth of Dominica, West Indies. Appl. Herpetol. **4**: 177-194.

McCranie, J.R., Wilson, L.D. (2002): The Amphibians of Honduras. Ithaca, New York, Society for the Study of Amphibians and Reptiles.

Perry, G., Gerber, G.P. (2006): Conservation of amphibians and reptiles in the British Virgin Islands: Status and patterns. Appl. Herpetol. **3**: 237-256.

Platenberg, R.J., Boulon, R.H., Jr. (2006): Conservation status of reptiles and amphibians in the U.S. Virgin Islands. Appl. Herpetol. **3**: 215-235.

Pounds, J.A., Bustamante, M.R., Coloma, L.A., Consuegra, J.A., Fogden, M.P.L., Foster, P.N., La Marca, E., Masters, K.L., Merino-Viteri, A., Puschendorf, R., Ron, S.R., Sanchez-Azofeifa, G.A., Still, C.J., Young, B.E. (2006): Widespread amphibian extinctions from epidemic disease driven by global warming. Nature **439**: 161-167.

Powell, R., 2006. Conservation of the herpetofauna on the Dutch Windward Islands: St. Eustatius, Saba, and St. Maarten. Appl. Herpetol. **3**: 293-306.

Pramuk, J.B. (2006): Phylogeny of South American *Bufo* (Anura: Bufonidae) inferred from combined evidence. Zool. J. Linn. Soc. **146**: 407-452.

Pramuk, J.B., Hass, C.A., Hedges, S.B. (2001): Molecular phylogeny and biogeography of West Indian toads (Anura: Bufonidae). Mol. Phylogenet. Evol. **20**: 294-301.

Savage, J.M. (2002): The Amphibians and Reptiles of Costa Rica. Chicago, University of Chicago Press.

Smith, M.L., Hedges, S.B., Buck, W., Hemphill, A., Inchaustegui, S., Ivie, M., Martina, D., Maunder, M., Ortega, J.F. (2005): Caribbean islands. In: Hotspots Revisited: Earth's Biologically Richest and Most Endangered Terrestrial Ecoregions, p. 112-118. Mittermeier, R.A., Gil, P.R., Hoffmann, M., Pilgrim, J., Brooks, T., Mittermeier, C.G., Lamoreux, J., da Fonseca, G.A., Eds, Mexico City, CEMEX.

Sodhi, N.S., Bickford, D., Diesmos, A.C., Lee, T.M., Koh, L.P., Brook, B.W., Sekercioglu, C.H., Bradshaw, C.J.A. (2008): Measuring the meltdown: Drivers of global amphibian extinction and decline. Plos One **3**: e1636.

Stuart, S.N., Chanson, J.S., Cox, N.A., Young, B.E., Rodrigues, A.S.L., Fischman, D.L., Waller, R.W. (2004): Status and trends of amphibian declines and extinctions worldwide. Science **306**: 1783-1786.

Viella, F.J. (1998): Biology of the mongoose (*Herpestes javanicus*) in a rain forest of Puerto Rico. Biotropica **30**: 120-125.

Voyles, J., Young, S., Berger, L., Campbell, C. Voyles, W.F., Dinudom, A., Cook, D., Webb, R., Alford, R.A., Skerratt, L.F., Speare, R. (2009): Pathogenesis of chytridiomycosis, a cause of catastrophic amphibian declines. Science **326**: 582-585.

Whitfield, S.M., Bell, K.E., Philippi, T., Sasa, M., Bolanos, F., Chaves, G., Savage, J.M., Donnelly, M.A. (2007): Amphibian and reptile declines over 35 years at La Selva, Costa Rica. Proc. Natl. Acad. Sci. USA **104**: 8352-8356.

Wilson, B.S., Horrocks, J.A., Hailey, A. (2006): Conservation of insular herpetofaunas in the West Indies Appl. Herpetol. **3**: 181-195.

Wilson, E.O. (1992): The Diversity of Life. Cambridge, Massachusetts, Harvard University Press.

Wilson, E.O. (2000): A global biodiversity map. Science **289**: 2279-2279.

Wilson, E.O. (2004): Taxonomy as a fundamental discipline. Philos. Trans. R. Soc. Lond. B Biol. Sci. **359**: 739.

Accepted: May 11, 2010 (AH).

Note added in proof

2010 updates for Cuba (observed by L.M. Díaz and A. Cádiz) and Hispaniola (observed by S.B. Hedges) are shown as * in table 1.

An overview of snake conservation in the West Indies

Peter J. Tolson[1], Robert W. Henderson[2]

[1] The Toledo Zoo, P.O. Box 140130, Toledo, OH 43614-0801, USA;
e-mail: ptolson@toledozoo.org
[2] Milwaukee Public Museum, Milwaukee, WI 53233-1478, USA

Abstract. No fewer than 120 snake species representing six families and 20 genera inhabit the West Indies; 115 (95.8%) are endemic to the region. Except for ± 30 taxa, we do not know the true conservation status of West Indian snakes; the herpetofauna is in a state of flux, as are the islands. Factors contributing to the decline of Antillean snake populations are complex, but nearly all are human-mediated and involve the introduction of exotic species, including predators (e.g., cats, black and Norway rats, mongooses) and ungulates (goats, pigs) that degrade habitats. Species that appear especially vulnerable to extirpations and extinctions are boids (*Epicrates* spp.) and diurnal, ground-dwelling colubrids (*Alsophis* spp. and *Liophis* spp.). Alterations in the prey base, commercial exploitation, and habitat destruction are likely responsible for declines in *Epicrates* populations, whereas predation by mongooses, cats, and rats have taken their toll on species of *Alsophis* and *Liophis*. Pesticides and herbicides may also have a deleterious impact on fossorial (e.g., *Typhlops* spp.) and anuran-eating species (*Antillophis* spp., *Chironius vincenti*, *Darlingtonia haetiana*). With greater environmental awareness and a minimum of tolerance, snakes and humans could co-exist.

Key words: *Alsophis*; cats; conservation; *Epicrates*; *Liophis*; mongoose; snakes; West Indies.

Introduction

The West Indies is one of the great centers of biodiversity on the planet (Myers et al., 2000; Smith et al., 2005). The snake fauna alone includes no fewer than 120 species (6 families, 20 genera; table 1), of which 115 (95.8%) are endemic to the slightly less than 220,000 km^2 that comprise the surface area of the region. Individual islands that harbor snakes may support anywhere from one to 30 species.

The snake fauna exhibits striking morphological and ecological diversity. Hispaniola, for example, supports three species of boids (*Epicrates*), three arboreal colubrids (*Uromacer*), two species of the enigmatic genus *Hypsirhynchus*, and nine species of fossorial blindsnakes (*Typhlops*). Cuba is home to the largest snake endemic to the region (*E. angulifer*, 4.0+ m) and a spectacular radiation of dwarf "boas" (*Tropidophis*, 15 species). The colubrid genus *Alsophis* is represented by 13

Table 1. Numbers of West Indian snake species by genus and their distribution. An asterisk (*) beside a number under an island or island group indicates that the genus has suffered one or more extirpations or extinctions on the island or island group. Genera having species that have suffered noticeable reductions in numbers on a particular island or island group are marked with a superscript 1.

Families	Bahamas	Cuba	Cayman Is.	Jamaica	Navassa	Hispaniola	Puerto Rico Bank	Lesser Antilles
Boidae								
Boa (1)	–	–	–	–	–	–	–	1
Corallus (2)	–	–	–	–	–	–	–	2[1]
Epicrates (9)	3[1]	1	–	1[1]	–	3	2*	–
Colubridae								
Alsophis (13)	1	1	1	1*	–	2*	3*	5*
Antillophis (2)	–	1	–	–	–	1	–	–
Arrhyton (12)	–	8	–	3	–	–	1	–
Chironius (1)	–	–	–	–	–	–	–	1
Clelia (2)	–	–	–	–	–	–	–	2*[1]
Darlingtonia (1)	–	–	–	–	–	1	–	–
Hypsirhynchus (2)	–	–	–	–	–	2	–	–
Ialtris (3)	–	–	–	–	–	3	–	–
Liophis (4)	–	–	–	–	–	–	–	4*
Mastigodryas (1)	–	–	–	–	–	–	–	1[1]
Nerodia (1)	–	1	–	–	–	–	–	–
Tretanorhinus (1)	–	1	1	–	–	–	–	–
Uromacer (3)	–	–	–	–	–	3	–	–
Leptotyphlopidae								
Leptotyphlops (7)	1	–	–	–	–	4	–	2
Tropidophiidae								
Tropidophis (27)	3	16	3	3	1*	1	–	–
Typhlopidae								
Typhlops (27)	3	2	2	1	1*	9	7	6
Viperidae								
Bothrops (2)	–	–	–	–	–	–	–	2

species occurring on more than 100 islands ranging from the Bahamas to Dominica. Similarly, 27 species of *Typhlops* are distributed from the Bahamas to Grenada, whereas 27 species of *Tropidophis* and nine species of *Epicrates* occur in the Bahamas and on the Greater Antilles. Diets of Antillean snakes include termites, anurans, lizards (primarily *Anolis*), birds, and mammals ranging in size from bats and introduced mice (*Mus musculus*) to native hutias (*Capromys, Geocapromys, Plagiodontia* spp.). Certain species (e.g., *Corallus grenadensis*, *Epicrates monensis*) may occur at population densities (40-125/ha) unheard of for snakes on the Neotropical mainland (Henderson, 2002; Tolson et al., in press).

This species-rich fauna is in great need of conservation. The arrival of Europeans in the West Indies in 1492 did much more than doom the local native American populations to extinction (Keegan, 1992); it set off a cascade of human-induced destructive events that have yet to reach a climax. The old-growth Bahamian and Antillean forests are gone, replaced at best by coppice and secondary forest and at worst by monocultures of exotic panic grass and agricultural fields. The vast majority of endemic terrestrial mammals, primarily hutias, that originally formed the prey base for the larger snakes are extinct, replaced by living human refuse such as the black and Norway rats (*Rattus rattus* and *R. norvegicus*), the house mouse (*Mus musculus*), the house cat (*Felis catus*), and the Indian mongoose (*Herpestes javanicus*).

Snakes are no more immune to the hazards of development and human overpopulation than the numerous other vertebrate species that have disappeared from the islands in the aftermath of human colonization. As with most vertebrates, the largest and most conspicuous species are generally the first to suffer, and many of the larger snakes of the West Indies, especially *Alsophis* and *Epicrates*, face an uncertain future in the wild. Herein we attempt to summarize the current status of selected West Indian snake species, the threats they face, and the life history attributes that make some more vulnerable to extinction than others.

Conservation Status of Antillean Snakes

Except for ca. 30 taxa (e.g. the boids and *Alsophis*), we know little about the actual status of many West Indian snake species. The difficulties associated with estimating tropical snake populations (e.g., difficulty of observation and remote chance of recapture) render that task daunting. Despite this dearth of knowledge regarding the true conservation status of most West Indian snakes, members of the genera *Alsophis* and *Liophis* have apparently fared the worst (table 1). At least two species, *A. ater* of Jamaica and *A. sancticrucis* of St. Croix, U.S. Virgin Islands are currently considered extinct (although *A. ater* is still listed as critically endangered on the IUCN Red List), and Hispaniolan *A. melanichnus* is probably also extinct. Several extinctions and extirpations of *Alsophis* have occurred in the Lesser Antilles as well (Henderson, 2004). *Liophis cursor* on Martinique may be extinct or, at best, restricted to 0.06 km^2 Rocher du Diamant; *L. ornatus* is found only on the St. Lucian satellite of Maria Major (0.09 km^2); and, on Barbados, *L. perfuscus* is likely extinct (Henderson, 2004). Likewise, Antillean *Epicrates*, especially the larger species, are the subjects of considerable international concern. Four taxa, *E. inornatus* of Puerto Rico (Endangered), *E. monensis monensis* of Isla Mona (Threatened), *E. m. granti* of Puerto Rico and the Virgin Islands (Endangered), and *E. subflavus* of Jamaica (Endangered), are currently listed by the U.S. Endangered Species Act. These three species also are given international protection under the Convention in Trade in Endangered Species (CITES, Appendix I). The Cuban boa, *Epicrates angulifer*, is considered to be an IUCN rare species. Although *Tropidophis bucculentus* from Navassa is presumably extinct (Powell, 1999), no species of *Tropidophis* is currently considered to be endangered, but this lack of legislative protection may be less a reflection of their true status than of our general lack of knowledge regarding the natural history of this genus.

Some species appear to be genuinely rare. Henderson (2004) discussed the extirpation and rarity of members of the colubrid genus *Clelia*, and Powell et al. (2000) commented on the scarcity of the colubrids *Alsophis anomalus*, *Ialtris agyrtes*, and *I. dorsalis* on Hispaniola. Henderson (2004) and Powell et al. (2000) suggested that these snakes may have always been rare. Nevertheless, all are ground-dwelling, thus rendering them susceptible to an array of introduced predators.

Factors Contributing to Declines

The factors adversely affecting populations of Antillean snakes are complex, but nearly all are human-mediated and involve the introduction of exotic species. We offer the following categories as the most serious threats.

Predation by exotic mammals

The waves of feral mammal introductions to the Antilles were certainly some of the most significant events in the decline and extirpation of the endemic West Indian fauna. The Indian mongoose, introduced to the Greater Antilles in 1872-1875 (Espeut, 1882; Hoagland et al., 1989), has been implicated in the extinction or extirpation of several species of *Alsophis* from major islands, including *A. ater* from Jamaica, probably *A. melanichnus* from Hispaniola, and *A. sancticrucis* from St. Croix, U.S. Virgin Islands (Henderson, 1992; Henderson and Sajdak, 1996). Also, although the ubiquitous *A. portoricensis* remains common in several areas of Puerto Rico, it can scarcely be found on either St. Thomas or St. John — the two largest U.S. Virgin Islands on the Puerto Rico Bank — and instead is restricted to offshore cays and satellite islands that are mongoose-free. Henderson (1992) noted that island size might be the critical factor in determining the effects that mongooses have on reptilian populations. However, several mongoose-free islands support *Alsophis* populations despite the presence of other exotic predators. As a terrestrial, diurnal predator, the mongoose undoubtedly represents the greatest danger to terrestrial diurnal snakes, although small, semi-fossorial species such as *Arrhyton* spp. may escape by virtue of their secretive habits, as Schmidt (1928) observed, referring to *Arrhyton exiguum*: "its secretive habits have preserved it from . . . the mongoose." Species of *Epicrates*, which are crepuscular and nocturnal and generally arboreal snakes (Tolson and Henderson, 1993), normally do not encounter this highly efficient predator unless basking on the ground or hiding in an easily accessible refuge.

Based on species that have suffered the most frequent extirpations and extinctions (i.e., species of *Alsophis* and *Liophis*; table 1), identifying certain characteristics that those species have in common is possible. They are diurnal, ground-dwelling, oviparous, often fast-moving and apparently susceptible to predation by ground-dwelling, diurnal, fast-moving mongooses. Species of *Alsophis* are ophidian counterparts of the lizard genus *Ameiva*; species of *Ameiva* are diurnal, ground-dwelling, fast-moving, oviparous, and populations have been dramatically reduced or eliminated from several *Herpestes*-infested islands in the Lesser Antilles (Powell and Henderson, 2005).

By far the most effective predator of nocturnal snakes is the house cat, *Felis catus*. As early as 1940, Lynn and Grant (1940) commented on the decline of *Epicrates subflavus* in the Portland Point area of Jamaica after the introduction of house cats. In 1991-1992, the Division of Wildlife, U.S. Virgin Islands rescued three individuals of *Epicrates m. granti* from house cats on St. Thomas, U.S. Virgin Islands (Tolson and Henderson, 1993). Tolson (1996) commented on cat predation on the Mona boa, and Tolson et al. (in press) documented an increase in the number of presumed cat-related injuries to Mona boas after Hurricane Georges in 1998. Both Wiewandt (1977) and García et al. (2001) documented predation on *A. portoricensis* on Isla Mona by feral house cats. Cats and dogs have been implicated in the extermination of *Tropidophis bucculentus* on Navassa (Powell, 1999).

Commensal species of introduced rodents, the black rat (*Rattus rattus*) and the Norway rat (*R. norvegicus*), are two of the most destructive predators in the West Indies. Rats are extremely adaptable in diet selection, can exist for long periods without water, and have an extremely high reproductive rate. Although approximately 95% of their diets consist of plant matter (Strecker et al., 1962), they are known to prey on invertebrates (Strecker et al., 1962), amphibians, reptiles (Whitaker, 1978), and birds (Atkinson, 1985; Austin, 1948; Johnstone, 1985). Because rats macerate their food so completely, detection of animal remains in rat stomachs is extremely difficult, and documentation of effects of rats on native faunas has usually come after invasions, with alarming declines noted in species numbers and diversity. Smaller species of snakes appear most vulnerable. Tolson (1988) found that *Epicrates m. granti* was absent in all sites surveyed on the Puerto Rico Bank ($n = 23$) that did not have an introduced rat predator, and, when *E. m. granti* was reintroduced to cays where rats had been eradicated, the boa populations flourished. Although many native predators, such as raptors, *Alsophis*, and the larger *Epicrates*, prey on rats (Henderson and Sajdak, 1996; Reagan, 1984; Rodríguez-Robles and Leal, 1993; Sheplan and Schwartz, 1974; Tolson and Henderson, 1993; Tolson et al., in press; Wiley, 2003), the rate of reproduction for rats is so high that population numbers easily keep pace with any losses due to predation.

Ungulates, particularly goats (*Capra hircus*) and pigs (*Sus scrofa*), indirectly affect snake populations by degrading habitat. Goats are capable of eliminating palatable vegetation from an area; several extensive expanses on West Indian islands that were heavily browsed by goats have degenerated into *Urochloa maxima* (= *Panicum maximum*) grassland, a habitat generally unsuited for *Epicrates* but certainly used by *Alsophis* in some instances. Repeated boa surveys in goat-browsed depression forests on Isla Mona failed to find evidence of *E. m. monensis*

(Tolson et al., in press). Pigs destroy native vegetation by rooting activities and will also consume snakes. Netting and Goin (1944) included an anecdotal account of the extirpation of *E. exsul* from Stranger's Cay, Little Bahama Bank after the introduction of pigs to the island.

Alterations in the prey base

Morgan and Woods (1986) and Woods (1989) estimated that 77-78 % of the original rodent fauna of the Antilles is now extinct. If the rate of predation of *E. angulifer* upon *Capromys pilorides* is any indication (where every adult *E. angulifer* examined to date has had hutia hair in its feces; PJT, unpub. observ.) of boid utilization of hutias as prey, consequences for other large snake species on islands where hutias have disappeared may have been considerable (e.g., *E. inornatus*, *E. subflavus*, and Bahamian *E. striatus*). Henderson et al. (1987) presented a likely historical scenario describing a transition from native rodents and insectivores that comprised the diet of *E. striatus* on pre-discovery Hispaniola to a diet of introduced *Mus* and *Rattus*. Similarly, prior to the arrival of Europeans and the introductions of *Mus* and *Rattus*, populations of *Corallus* in the Lesser Antilles probably fed on species of murid rodents (Oryzomyini, probably *Oryzomys*; based on remains from archaeological sites: Lippold, 1991; Pregill et al., 1994; Henderson et al., 1996).

Rock iguana populations of the genus *Cyclura* have shown catastrophic declines on many islands (see Alberts, 2000) yet remain important prey items in the diet of both *Alsophis* and *Epicrates* (Knapp and Owens, 2004; Tolson et al., in press). The impact of the decline in *Cyclura* populations is unknown, but the loss of *C. pinguis* from the Puerto Rico Bank may be mitigated somewhat by the introduction of *Iguana iguana*, which is heavily utilized as prey by *Epicrates* (PJT, unpub. observ.).

Commercial exploitation

Although international trade for all boids is currently regulated by CITES, substantial numbers of *Epicrates* and some *Tropidophis* are collected annually for the pet trade. Dodd (1986) reported that 4752 individuals of ten insular species of *Epicrates* and *Tropidophis* were imported into the United States from 1977-1983. These numbers probably represent only a fraction of these snakes actually entering the country. For example, during this eight-year period, only two individuals of *E. chrysogaster* are reported as entering the United States, yet commercial dealers offered hundreds of these animals for sale during the same period. The island of South Bimini once had an extremely dense population of *E. striatus fosteri*. These animals were removed illegally from the island by the hundreds during the 1970s for the pet trade (Tolson and Henderson, 1993).

Habitat destruction

The sad litany of forest destruction in the West Indies has been voiced many times (Asprey and Robbins, 1953; Beard, 1949; Harcourt and Ottenwalder, 1996; Wadsworth, 1950), yet we doubt that any of us can appreciate the magnitude of destruction and its consequences for wildlife. Myers et al. (2000) estimated that only 11.3% of the original vegetation in the Caribbean region remains. How these floristic changes have impacted snakes is unclear, as at least some of these species are extremely adaptable. For example, although *Epicrates angulifer* was primarily considered to be a forest species (Sheplan and Schwartz, 1974), we have found it to be very common in highly disturbed areas with adequate grassland cover. Similarly, some of the highest concentrations of Antillean snakes we have ever observed, including *Antillophis parvifrons*, *Epicrates fordi*, *E. s. striatus*, *Hypsirhynchus ferox*, and *Uromacer* spp. have been in the vicinity of the Haitian villages of Limbė and Thomazeau, where little original vegetation remained. Both Sifers et al. (2001) and Powell et al. (1996) have demonstrated the abundance of certain *Anolis* spp. in human-altered edge habitats as opposed to forest interiors, and in numerous field trips we have found *Epicrates chrysogaster*, *E. exsul*, *E. m. granti*, *E. striatus*, *Corallus cookii*, and *C. grenadensis* to be more common in these situations than in the interiors of woodlands. Other species, however, such as *E. inornatus*, *E. subflavus*, and *Chironius vincenti* (Henderson and Hass, 1993) seem to prefer more heavily forested situations, and the widespread deforestation of Jamaica and Puerto Rico has been implicated in the decline of both species of *Epicrates* (Oliver, 1982; USFWS, 1986). Both of these species have reportedly become more numerous as secondary forests have become more mature and new areas are surveyed (Dávalos and Eriksson, 2004; Wiley, 2003; Wunderle et al., 2004). The most serious consequences seem to accrue when environmental insults follow in close succession, such as the deforestation of the Virgin Islands that was followed by large-scale annual burning that accompanied the cultivation of sugar cane. Such activities may explain why *E. m. granti* is limited to the east end of St. Thomas (Nellis et al., 1983), where sugar cane cultivation was secondary to cattle ranching. Reduction in the numbers of treeboas (*Corallus grenadensis*) encountered at several sites on Grenada has been documented, but the proximal causes of those diminished numbers have not been determined (Henderson and Berg, 2006). Although 2004's Hurricane Ivan severely altered *C. grenadensis* habitat over much of the island, declining encounter rates were observed prior to the damage.

Discussion

The West Indian herpetofauna is in a state of flux, as are the islands themselves. The introduction of predators, prey species, and potential competitors in concert with dramatic alterations of habitat has produced fragmented or constricted distributions and altered ecologies. The past 40 years have witnessed a decrease in the proportion of humans living in rural areas of Latin America and the Caribbean from about one-half to less than one-quarter (Aide and Grau, 2004). Although this may bode well for snake species removed from urban centers, dramatic and far-reaching changes have already been wrought on most West Indian islands and their habitats.

As noted above, more often than not we know little about the real conservation status of West Indian snakes. Deforestation and introduced vertebrates are blatant and conspicuous threats to native wildlife. Many abuses of the environment, however, may not be visible, and the impact on island snake faunas may not be detected or understood for many years. For example, we cannot yet assess the long-term impact of pesticides and herbicides on fossorial species (*Leptotyphlops*, *Typhlops*) in active and abandoned agricultural areas. Scolecophidian species are secretive and not usually encountered unless an active, focused search is conducted. Similarly, herbicides and pesticides may adversely affect anuran populations that, in turn, have a deleterious impact on frog-eating snakes (e.g., *Antillophis parvifrons*, *Darlingtonia haetiana*, *Chironius vincenti*).

Larger snakes do exist in high densities even in relatively disturbed habitat. Examples include *A. antiguae* on Great Bird Island (Antigua) (Daltry et al., 2001), *A. portoricensis* on the outlying cays of the Virgin Islands (e.g., Barun et al., in press), *E. angulifer* at the Cueva de las Majas near Trinidad, Cuba (Hardy, 1957), *E. inornatus* at the Cueva de los Culebrónes in the vicinity of Arecibo (Rodríguez and Reagan, 1984), *E. monensis* on Isla Mona (Tolson et al., in press), *E. striatus* in the vicinity of Limbe, Haiti (Tolson and Henderson, 1993), and *Corallus grenadensis* at several sites on Grenada (Henderson, 2002). Given that several different species can exist in high densities, their general absence from areas of apparently suitable habitat can be construed as evidence of declines.

The many anecdotal reports in the literature of snake population declines in the West Indies, especially those of Barbour (e.g., 1914, 1937), indicate that some of these accounts need to be taken seriously (although some of Barbour's assessments are based on little or no personal experience; Henderson and Powell, 2004). In order to document changes in snake numbers, document population fluctuations, or determine responses to habitat alterations, no alternative replaces long-term field studies. Although we do not discourage studies in what relatively pristine situations yet exist, we do encourage biologists to study snakes in sub-optimal situations as well, in order to determine how particular species are responding to the ecological glitches perpetrated by the arrival and persistent activities of humans (e.g., Powell et al., 1996).

Although snake-human encounters likely often end with the snake being killed with a machete or rock, we would like to believe that, with minimal tolerance and greater awareness of a given island's natural history, snakes and humans can co-occur. On Guana Island in the British Virgin Islands, *Alsophis portoricensis* is common throughout the island, including a resort complex. The killing of snakes is forbidden, and the snakes sun and forage for anoles on and adjacent to pedestrian walkways and occasionally enter buildings. At night, the artificially illuminated walkways attract insects that, in turn, attract diurnally active anoles and, occasionally, a diurnal *Alsophis* will nocturnally forage for anoles at the electric lights (Perry and Lazell, 2000). Thus, in the absence of active persecution, some snake species are able to survive (perhaps even thrive) in close proximity to humans (Barun et al., in press; Powell and Henderson, 2006). With only two species (1.7% of the entire snake fauna) of dangerously venomous snakes in the entire region, no valid reason exists why snakes should not be afforded the same respect, admiration, and protection as high profile species with feathers or fur.

Acknowledgements. Henderson is grateful to the Milwaukee Public Museum, the Windway Foundation, Milwaukee Country Zoo, the Milwaukee Zoological Society, and the National Science Foundation (Grant No. DBI-9732257 to Robert Powell) for recent funding of West Indian fieldwork. The enthusiastic participation of Craig Berg, Ky Henderson, Bob Powell, and Rich Sajdak is most appreciated. Tolson is grateful to The Toledo Zoo, the U.S. Fish and Wildlife Service, the Departamento de Recursos Naturales y Ambientales de Puerto Rico, the Division of Wildlife, U.S. Virgin Islands, The Institute of Museum and Library Services, and the U.S. Navy, all of which contributed the significant financial and logistical support that enabled long-term studies of *Epicrates*. He treasures the camaraderie and dedication of the many friends and colleagues that are fellow-travelers in snake conservation: Miguel A. García, Judy Pierce, Candee L. Ellsworth, Alberto O. Alvarez, and Carlos E. Díez.

References

Aide, T.M., Grau, H.R. (2004): Globalization, migration, and Latin American ecosystems. Science **305**: 1915-1916.

Alberts, A.C., ed. (2000): West Indian Iguanas: Status Survey and Conservation Action Plan. Gland, Switzerland, IUCN-the World Conservation Union.

Asprey, G.F., Robbins, R.G. (1953): The vegetation of Jamaica. Ecol. Mono. **23**: 359-412.

Atkinson, I.A.E. (1985): The spread of commensal species of *Rattus* to oceanic islands, and their effects on island avifaunas. In: Conservation of Island Birds, p. 35-81. Moors, P.J., Ed., ICBP Tech. Publ. No. 3, Cambridge, England.

Austin, O.L. (1948): Predation by the common rat, *Rattus norvegicus*, in the Cape Cod colonies of nesting terns. Bird Band. **14**: 60-65.

Barbour, T. (1914): A contribution to the zoogeography of the West Indies, with especial reference to amphibians and reptiles. Mem. Mus. Comp. Zool. **44**: 1-359.

Barbour, T. (1937): Third list of Antillean reptiles and amphibians. Bull. Mus. Comp. Zool. **82**: 77-136.

Barun, A., Perry, G., Henderson, R.W., Powell, R. (in press). *Alsophis portoricensis anegadae* (Squamata: Colubridae): morphometric characteristics, activity patterns, and habitat use. Copeia.

Beard, J.S. (1949): Natural vegetation of the Windward and Leeward islands. Oxford Forestry Mem. 21.

Daltry, J.C., Cooper, G., Day, M.L., Harper, J., Henry, M., Lindsay, K., Smith, B.E. (2001): Five years of conserving the "world's rarest snake", the Antiguan Racer, *Alsophis antiguae*. Oryx **35**: 119-127.

Dávalos, L.M., Eriksson, E. (2004): *Epicrates subflavus* (Jamaican boa). Foraging Behavior. Herp. Rev. **35**: 66.

Dodd, C.K., Jr. (1986): Importation of live snakes and snake products into the United States, 1977-1983. Herpetol. Rev. **17**: 76-79.

Espeut W.B. (1882): On the acclimatization of the Indian Mungoos in Jamaica. Proc. Zool. Soc. London, **1882**: 712-714.

García, M.A., Díez, C.E., Alvarez, A.O. (2001): The impact of feral cats on Mona Island wildlife and recommendations for their control. Carib. J. Sci. **37**: 107-108.

Harcourt, C., Ottenwalder, J.A. (1996): Hispaniola. In: The Conservation Atlas of Tropical Forests, the Americas, p. 102-111. Harcourt, C.S., Sayer, J.A., Eds, Gland, Switzerland, IUCN.

Hardy, J.D. (1957): Bat predation by the Cuban boa, *Epicrates angulifer* Bibron. Copeia **1957**: 151-152.

Henderson, R.W. (1992): Consequences of predator introductions and habitat destruction on amphibians and reptiles in the post-Columbus West Indies. Carib. J. Sci. **28**: 1-10.

Henderson, R.W. (2002): Neotropical Treeboas: Natural History of the *Corallus hortulanus* Complex. Malabar, Florida, Krieger Publ. Co.

Henderson, R.W. (2004): Lesser Antillean snake faunas: distribution, ecology, and conservation concerns. Oryx **38**: 311-320.

Henderson, R.W., Berg, C.S. (2006): The herpetofauna of Grenada and the Grenada Grenadines: conservation concerns. Applied Herpetol. **3**: 197-213.

Henderson, R.W., Haas, G.T. (1993): Status of the West Indian snake *Chironius vincenti*. Oryx **27**: 181-184.

Henderson, R.W., Noeske-Hallin, T.A., Ottenwalder, J.A., Schwartz, A. (1987): On the diet of the boa *Epicrates striatus* on Hispaniola, with notes on *E. fordi* and *E. gracilis*. Amphibia-Reptilia **8**: 251-258.

Henderson, R.W., Powell, R. (2004): Thomas Barbour and the *Utowana* voyages (1929-1934) in the West Indies. Bonn. Zool. Beit. **52**: 297-309.

Henderson, R.W., Sajdak, R.A. (1996): Diets of West Indian racers (Colubridae: *Alsophis*): composition and biogeographic implications. In: Contributions to West Indian herpetology: a tribute to Albert Schwartz, p. 327-338. Powell, R., Henderson, R.W., Eds, Ithaca, Soc. Study Amphibians and Reptiles, Contrib. Herpetol. 12.

Henderson, R.W., Winstel, R.A., Friesch, J. (1996): *Corallus hortulanus* (Serpentes: Boidae) in the post-Columbian West Indies: new habitats, new prey species, and new predators. In: Contributions to West Indian Herpetology: A Tribute to Albert Schwartz, p. 417-423. Powell, R., Henderson, R.W., Eds, Ithaca, Soc. Study Amphibians and Reptiles, Contrib. Herpetol. 12.

Hoagland, D.B., Horst, G.R., Kilpatrick, C.W. (1989): Biogeography and population biology of the mongoose in the West Indies. In: Biogeography of the West Indies: Past, Present, and Future, p. 611-633. Woods, C.A., Ed., Gainesville, Sandhill Crane Press.

Johnstone, G.W. (1985): Threats to birds on subantarctic islands. In: Conservation of Island Birds, p. 101-121. Moors, P.J., Ed., ICBP Tech. Publ. No. 3, Cambridge, England.

Keegan, W.F. (1992): Destruction of the Taino. Archaeology **45**: 51-56.

Knapp, C.R., Owens, A.K. (2004): Diurnal refugia and novel ecological attributes of the Bahamian boa, *Epicrates striatus fowleri* (Boidae). Carib. J. Sci. **40**: 265-270.

Lippold, L.K. (1991): Animal resource utilization by Saladoid peoples at Pearls, Grenada, West Indies. In: Proceedings of the 13th International Congress for Caribbean Archaeology, p. 264-268. Ayubi, E.N., Haviser, J.B., Eds, Netherlands Antilles, Rept. Archaeol. Anthropol. Inst., No. 9.

Lynn, W.G., Grant, C. (1940): The herpetology of Jamaica. Bull. Inst. Jamaica, Sci. Ser. 1.

Morgan, G.S., Woods, C.A. (1986): Extinction and the zoogeography of West Indian land mammals. Biol. Linn. Soc. **28**: 167-203.

Myers, N., Mittermeier, R.A., Mittermeier, C.G., de Fonseca, G.A., Ke, J. (2000): Biodiversity hotspots for conservation priorities. Nature **403**: 853-858.

Nellis, D.W., Norton, R.L., MacLean, W.P. (1983): On the biogeography of the Virgin Islands boa, *Epicrates monensis granti*. J. Herpetol. **17**: 413-417.

Netting, M.G., Goin, C.J. (1944): Another new boa of the genus *Epicrates* from the Bahamas. Ann. Carnegie Mus. **30**: 71-76.

Oliver, W.L.R. (1982): The coney and the yellow snake: The distribution and status of the Jamaican hutia *Geocapromys browni* and the Jamaican boa *Epicrates subflavus*. Dodo, J. Jersey Wildl. Preserv. Trust **19**: 6-33.

Perry, G., Lazell, J. (2000): *Liophis portoricensis anegadae* (NCN). Night-light niche. Herpetol. Rev. **31**: 247.

Powell, R. (1999): Herpetology of Navassa Island, West Indies. Carib. J. Sci. **35**: 1-13.

Powell, R., Henderson, R.W. (2005): Conservation status of Lesser Antillean reptiles. Iguana **12**: 2-17.

Powell, R., Henderson, R.W. (2006): Urban herpetology in the West Indies. In: Urban Herpetology. Jung, R.E., Mitchell, J.C., Eds, Salt Lake City, Soc. Study Amphibians and Reptiles, Herpetol. Conserv. Vol. 3.

Powell, R., Parmerlee, Jr., J.S., Smith, D.D. (1996): Evidence of spatial niche partitioning by a Hispaniolan lizard community in a xeric habitat. In: Contributions to West Indian Herpetology: A Tribute to Albert Schwartz, p. 317-326. Powell, R., Henderson, R.W., Eds, Ithaca, Soc. Study Amphibians and Reptiles, Contrib. Herpetol. 12.

Powell, R., Ottenwalder, J.A., Incháustegui, S.J., Henderson, R.W., Glor, R.E. (2000): Amphibians and reptiles of the Dominican Republic: Species of special concern. Oryx **34**: 118-128.

Pregill, G.K., Steadman, D.W., Watters, D.R. (1994): Late Quaternary vertebrate faunas of the Lesser Antilles: Historical components of Caribbean biogeography. Bull. Carnegies Mus. Nat. Hist. **30**: iii, 1-51.

Reagan, D.P. (1984): Ecology of the Puerto Rican boa (*Epicrates inornatus*) in the Luquillo Mountain of Puerto Rico. Carib. J. Sci. **20**: 119-126.

Rodrígues-Robles, J.A., Leal, M. (1993): *Alsophis portoricensis* (Puerto Rican Racer) diet. Herpetol. Rev. **24**: 150-151.

Rodríguez, G., Reagan, D.P. (1984): Bat predation by the Puerto Rican boa (*Epicrates inornatus*). Copeia **1984**: 219-220.

Schmidt, K.P. (1928): Amphibians and land reptiles of Porto Rico, with a list of those reported from the Virgin Islands. New York Acad. Sci., Scientific Survey of Porto Rico and the Virgin Islands, vol. X, pt. 160 pp.

Sheplan, B.R., Schwartz, A. (1974): Hispaniolan boas of the genus *Epicrates* (Serpentes, Boidae) and their Antillean relationships. Ann. Carnegie Mus. **45**: 57-143.

Sifers, S.M., Yeska, M.L., Ramos, Y.M., Powell, R., Parmerlee, J.S., Jr. (2001): *Anolis* lizards restricted to altered edge habitats in a Hispaniolan cloud forest. Carib. J. Sci. **37**: 55-62.

Smith, M.L., Hedges, S.B., Buck, W., Hemphill, A., Incháustegui, S., Ivie, M., Martina, D., Maunder, M., Ortech, J.F. (2005): Caribbean Islands. In: Hotspots Revisited: Earth's Biologically Richest and Most Endangered Terrestrial Ecoregions, p. 112-118. Mittermeier, R.A., Gill, P.R., Hoffman, M., Pilgrim, J., Brooks, T., Mittermeier, C.G., Lamoreaux, J., da Fonseca, G.A.B., Eds, Mexico City, CEMEX.

Strecker, R.L., Marshall, J.T., Jr., Jackson, W.B., Barbehenn, K.R., Johnson, D.H. (1962): Pacific island rat ecology. Bernice P. Bishop Mus. Bull. **225**: 1-274.

Tolson, P.J. (1996): Conservation of *Epicrates monensis* on the satellite islands of Puerto Rico. In: Contributions to West Indian Herpetology: A Tribute to Albert Schwartz, p. 407-416. Powell, R., Henderson, R.W., Eds, Soc. Study Amphib. Rept., Ithaca, New York. Contrib. Herpetol., Vol. 12.

Tolson, P.J., Garcia, M.A., Ellsworth, C.L. (in press): Habitat Use by the Mona Boa (*Epicrates m. monensis*) on Isla Mona, West Indies. In: Biology of Boas and Pythons. Henderson, R.W., Powell, R., Eds, Eagle Mountain, Utah, Eagle Mountain Publ.

Tolson, P.J., Henderson, R.W. (1993): The Natural History of West Indian Boas. Taunton, England, R & A Publ., Ltd.

USFWS (U.S. Fish and Wildlife Service) (1986): Puerto Rican Boa Recovery Plan. U.S. Fish and Wildlife Service, Atlanta, Georgia. 21 p.

Wadsworth, F.H. (1950): Notes on the climax forests of Puerto Rico and their destruction and conservation prior to 1900. Carib. Forester **11**: 8-47.

Wiewandt, T.A. (1977): Ecology, behavior, and management of the Mona Island Ground Iguana, *Cyclura stejnegeri*. Unpubl. Ph.D. Dissertation, Cornell Univ. 338 pp.

Whitaker, A.H. (1978): The effects of rodents on reptiles and amphibians. In: The Ecology and Control of Rodents in New Zealand Nature Reserves, p. 75-86. Dingwall, P.R., Atkinson, I.A.E., Hay, C., Eds, New Zealand Dept. Lands and Surv. Info. Ser. No. 4.

Wiley, J.W. (2003): Habitat association, size, stomach contents, and reproductive condition of Puerto Rican Boas (*Epicrates inornatus*). Carib. J. Sci. **39**: 189-194.

Woods, C.A. (1989): Endemic rodents of the West Indies: The end of a splendid isolation. Pp. 11-19. In: Rodents. A World Survey of Conservation Concern, p. 11-19. Lidiker, W., Jr., Ed., Occ. Pap. IUCN SSC No. 4.

Wunderle, J.M., Mercado, J.E., Parresol, B., Terranova, E. (2004): Spatial ecology of Puerto Rican Boas (*Epicrates inornatus*) in a hurricane impacted forest. Biotropica **36**: 555-571.

Accepted: July 3, 2006 (BSW).
Reprinted from Applied Herpetology **3**: 345-356 (2006).

Addendum

Many changes in the classification of West Indian snakes have occurred since 2006. Most notable is partitioning of several genera formerly placed in the Colubridae but now considered members of the Dipsadidae (see Hedges et al. 2009, and Zaher et al., 2009 for details). Likewise, West Indian *Leptotyphlops* (family Leptotyphlopidae) were partitioned into several genera (see Adalsteinsson et al., 2009). Table A1 summarizes those changes and updates table 1 in the paper. About 150 species are now recognized.

In addition to changes in classification, four snake species have been introduced to the West Indies, almost certainly via human agency, and current evidence indicates that reproductive populations have been established on some islands. They are the colubrids *Pantherophis guttatus* (Turks and Caicos Islands, Grand Cayman, Virgin Islands, and Lesser Antilles) and *Tantilla melanocephala* (Grenada Bank), the natricid *Storeria dekayi* (Bahamas), and the typhlopid *Ramphotyphlops braminus* (Turks and Caicos, Grand Cayman, and Lesser Antilles). *Tantilla melanocephala*, *S. dekayi*, and *R. braminus* are invertebrate predators and they are unlikely to represent a competitive threat to snake species native to the islands on which they have been introduced. *Pantherophis guttatus*, however, is a small-mammal predator and could compete with vertebrate-eating snakes on the islands where it is established.

References

Adalsteinsson, S.A., Branch, W.R., Trape, S., Vitt, L.J., Hedges, S.B. (2009): Molecular phylogeny, classification, and biogeography of snakes of the family Leptotyphlopidae (Reptilia, Squamata). Zootaxa **2244**: 1-50.

Hedges, S.B., Couloux, A., Vidal, N. (2009): Molecular phylogeny, classification, and biogeography of West Indian racer snakes of the Tribe Alsophiini (Squamata, Dipsadidae, Xenodontinae). Zootaxa **2067**: 1-28.

Zaher, H., Grazziotin, F.G., Cadle, J.E., Murphy, R.W., de Moura-Leite, J.C., Bonatto, S.L. (2009): Molecular phylogeny of advanced snakes (Serpentes, Caenophidia) with an emphasis on South American xenodontines: A revised classification and descriptions of new taxa. Pap. Avul. Zool. **49**: 115-153.

Table A1. Numbers of West Indian snake species by genus and island or island group. An underlined genus name indicates that species in that genus were introduced via human agency. An asterisk (*) beside a number under an island or island group indicates that the genus has suffered one or more extirpations or extinctions on the island or island group. Genera having species that have suffered noticeable reductions in numbers on a particular island or island group are marked with a superscript plus sign (+).

Family Genus (No. species)	Bahamas[a]	Cuba	Cayman Islands	Jamaica	Navassa	Hispaniola	Puerto Rico Bank[b]	Lesser Antilles
Boidae (13)								
Boa (2)	–	–	–	–	–	–	–	2
Corallus (2)	–	–	–	–	–	–	–	2^{+}
Epicrates (9)	3^{+}	1	–	1^{+}	–	3	2*	–
Colubridae (4)								
Chironius (1)	–	–	–	–	–	–	–	1
Mastigodryas (1)	–	–	–	–	–	–	–	1^{+}
Pantherophis (1)	–	–	1	–	–	–	1	1
Tantilla (1)	–	–	–	–	–	–	–	1
Dipsadidae (50)								
Alsophis (8)	–	–	–	–	–	–	–	8*
Arrhyton (8)	–	8	–	–	–	–	–	–
Borikenophis (3)[c]	–	–	–	–	–	–	3*	–
Caraiba (1)[d]	–	1	–	–	–	–	–	–
Clelia (2)	–	–	–	–	–	–	–	2^{*+}
Cubophis (5)[c]	1	1	3	–	–	–	–	–
Haitiophis (1)[e]	–	–	–	–	–	1	–	–
Hypsirhynchus (8)[f]	–	–	–	4*	–	4*	–	–
Ialtris (4)	–	–	–	–	–	4	–	–
Liophis (4)	–	–	–	–	–	–	–	4*
Magliophis (2)[g]	–	–	–	–	–	–	2	–
Tretanorhinus (1)	–	1	1	–	–	–	–	–
Uromacer (3)	–	–	–	–	–	3	–	–

Table A1. (Continued).

Family Genus (No. species)	Bahamas[a]	Cuba	Cayman Islands	Jamaica	Navassa	Hispaniola	Puerto Rico Bank[b]	Lesser Antilles
Leptotyphlopidae (9)								
Epictia (2)[h,i]	1	–	–	–	–	–	–	–
Mitophis (4)[i]	–	–	–	–	–	4	–	–
Tetracheilostoma (3)[i]	–	–	–	–	–	–	–	3
Natricidae (2)								
Nerodia (1)	–	1	–	–	–	–	–	–
Storeria (1)	1	–	–	–	–	–	–	–
Tropidophiidae (27)								
Tropidophis (27)	2	16	3	3	1*	1	–	–
Typhlopidae (42)								
Ramphotyphlops (1)	1	–	1	–	–	–	–	1
Typhlops (41)	3	9	2	1	1*	13	8	6
Viperidae (2)								
Bothrops (2)	–	–	–	–	–	–	–	2

[a] Including the Turks and Caicos Islands.
[b] Plus Isla Mona, St. Croix, and associated islets.
[c] Includes species formerly assigned to the genus *Alsophis*.
[d] Formerly *Antillophis andreae*.
[e] Formerly *Alsophis anomalus*.
[f] Includes species formerly in the genera *Alsophis*, *Antillophis*, *Arrhyton*, and *Darlingtonia*.
[g] Includes species formerly in the genus *Arrhyton*.
[h] The second species occurs on the Swan Islands.
[i] Includes species formerly in the genus *Leptotyphlops*.

Introduced amphibians and reptiles in the greater Caribbean: Patterns and conservation implications

Robert Powell[1], Robert W. Henderson[2], Michael C. Farmer[3], Michel Breuil[4], Arthur C. Echternacht[5], Gerard van Buurt[6], Christina M. Romagosa[7], Gad Perry[8,9]

[1] Department of Biology, Avila University, Kansas City, MO 64145, USA
[2] Section of Vertebrate Zoology, Milwaukee Public Museum, Milwaukee, WI 53233, USA
[3] Department of Agricultural and Applied Economics, Texas Tech University, Box 42132, Lubbock, TX 79409, USA
[4] Département de Systématique et d'Évolution, Taxonomie et Collections, Reptiles et Amphibiens, Muséum national d'Histoire naturelle, 75005 Paris, France
[5] Department of Ecology and Evolutionary Biology, University of Tennessee, Knoxville, TN 37996, USA
[6] Kaya Oy Sprock 18, Curaçao
[7] Department of Biological Sciences, Auburn University, Auburn, AL 36849, USA
[8] Department of Natural Resource Management, Texas Tech University, Lubbock, TX 79409, USA
[9] Corresponding author; e-mail: Gad.Perry@ttu.edu

Abstract. Non-native species are a growing worldwide conservation problem, often second only to habitat destruction and alteration as a cause of extirpations and extinctions. Introduced taxa affect native faunas through competition, predation, hybridization, transmission of diseases, and even by confounding conservation efforts focused on superficially similar endemic species. The number of known introductions of amphibian and reptilian species continues to grow. Herein, we document the arrival and establishment of alien amphibians and reptiles in the greater Caribbean region and the means by which they arrived. These include 130 species (25 amphibians and 105 reptiles) responsible for 364 individual introductions, of which 70.3% resulted in populations established for at least a short period. The impact of those 256 established populations ranges from minimal (localized effects largely restricted to dramatically altered habitats) to severe (displacement of native species from natural and modified habitats). Although intentional introductions for putative pest control (mostly historical) and food (historical and ongoing) are factors in some instances, the primary pathways for introductions today are inadvertent. Nearly all are associated with either the ever-growing pet trade or stowaways in cargo and ornamental plants. To document the extent of the live animal trade for pets and food, we review the surprisingly large numbers of documented individuals exported from the Caribbean into the United States (US) and from the US to the Caribbean. The extent of such trade and the rates of non-native arrivals continue to increase, and both are related to indices of regional economic activity. Because prevention is by far better — and more economical — than eradication of an established alien, we recommend increased scrutiny of transported goods and animals to and from the islands. An integrated policy response is clearly necessary to address what is a regional issue. Although the region is highly fragmented both geographically and politically, we urge an increased regional cooperation for fighting invasive species in general and invasive herpetofauna in particular. Precedents for such cooperation include the Caribbean Community and Common Market (CARICOM) and the Caribbean Cooperation in Health initiative.

Key words: Amphibians; Caribbean; economic activity; eradication; introduced species; live animal trade; prevention; regional cooperation; reptiles; urban; vectors.

Introduction

Natural dispersal is a common phenomenon, although long-distance dispersal is typically infrequent (Nathan et al., 2003; Trakhtenbrot et al., 2005). Human-aided dispersal is increasingly common, however, even over great distances. Globally, human-transported non-native species are among the top three causes of biodiversity loss (Clavero and García-Berthou, 2005; McGeoch et al., 2010). The number of amphibians and reptiles being moved to non-native locations is growing (Lever, 2003; Kraus, 2009), as are reports of their ecological and economic impacts (e.g., Bomford et al., 2009), despite the inadequate attention paid to documenting them (McGeoch et al., 2010).

The greater Caribbean region, with extensive tourism in many areas and limited local production of essential items such as food and building materials, is at especially high risk. Herpetological introductions in the region are not new. Félix-Louis L'Herminier, as director of the "Jardin de naturalization de la Guadeloupe" in the early 19th century, had a goal of introducing and acclimating new species to the island (Breuil, 2002, 2003). Among the species he attempted to establish were *Kinixys erosa*, *Kinixys homeana*, and *Pelusios castaneus*, which are native to western Africa and which he might have purchased from slave traders. In addition, his son, François-Joseph, visited Puerto Rico and caught *Trachemys stejnegeri*, which was liberated in Marie-Galante. Other 19th-century reports include Schomburgk (1848), Gosse (1851), Feilden (1889), and Boulenger (1891). Modern reports are numerous and highly dispersed, despite efforts of Lever (2003) and Kraus (2009) to collate them. Our goal in this chapter is to summarize what is known about herpetological introductions in the region, the mechanisms that allow them, and their effects in this wide geographical area.

By their nature, islands are more isolated than mainland sites, yet over-water dispersal still occurs naturally (e.g., Censky et al., 1998; Calsbeek and Smith, 2003). We exclude such instances from the current analysis, which focuses on human-aided extra-limital dispersal events. We hope that the broad patterns that emerge — in particular, the primacy of a small number of arrival mechanisms and the close relation with economic activity — will encourage a coordinated regional policy response and help reduce negative economic and ecological impacts.

Materials and Methods

Regional coverage

We define the "greater Caribbean" to include the West Indies biogeographic region, the oceanic islands of Isla de San Andres and Isla de Providencia (Colombia), and three continental islands off the northern shore of South America (Aruba, Bonaire, Curaçao). For the West Indies, we use the definition of Henderson and Powell (2009) to include the Greater and Lesser Antilles plus the Bahama, Turks and Caicos, Swan, and Cayman islands. We exclude continental islands that have been connected to the mainland until recently and with faunas that reflect that origin. These include Trinidad, Isla de Margarita off South America (SA), and the Honduran Bay Islands off the Central American (CA) coast. We also exclude Tobago, which is functionally a continental island due to its proximity to Trinidad.

Literature review

To develop an overview of all introductions of amphibians or reptiles in the region, we exhaustively reviewed the pertinent literature, much of which was reviewed previously in Kraus (2009) and Henderson and Powell (2009). Unfortunately, records of "benign" non-native arrivals and dispersal are notoriously incomplete (McGeoch et al., 2010). We therefore supplemented the literature accounts with our own personal experiences, collected over several decades of working in the region. Finally, we solicited supplementary information from persons in parts of the region for which data were sparse.

We organize our text taxonomically. Written accounts identify (when known), the arrival mechanism (often as identified in Kraus, 2009), and whether this was a one-time arrival, a repeated incursion, or an established population. However, the origin of some populations — whether they arrived naturally or were human-mediated — remains uncertain. Locations are detailed in appendices 1 and 2, which also provide citations to assist readers seeking information regarding the sources or fates of introductions unrelated (appendix 1) and related to (appendix 2) conservation and research efforts. To avoid unnecessary duplication, we do not consistently distinguish arrivals to single islands within island groups or banks (e.g., Bahamas, Virgin Islands, Grenadines, Guadeloupean Archipelago) from arrivals to an entire island group.

Not all introductions are successful. Reports of one-time arrivals (e.g., Powell et al., 2005; Perry, 2009a) are uncommon in the literature, although they provide valuable information on vectors, propagule pressure, and times of arrival. In some instances, we report the presence of ephemeral populations, although many lasted for only relatively short periods. For example, Powell et al. (1992) documented a population of *Anolis bimaculatus* on St. Maarten that included both adults and juveniles, presumably from St. Eustatius. Subsequent visits to the site where the original observations were made and to nearby areas with presumably ideal habitat failed to reveal additional individuals. When known, we indicate such outcomes.

However, some populations indicated as established may yet fail, and some failures almost certainly have gone undocumented.

The source of introduced populations is only sometimes known, even when the event was recent. Generally, we are even less certain of sources for older introductions. For example, Amerindians and early European colonists almost certainly intentionally transported tortoises (*Chelonoidis carbonaria*) and iguanas (*Iguana iguana*) and possibly rainfrogs (*Eleutherodactylus johnstonei*, albeit inadvertently), from the mainland to islands or from one island to another (e.g., Censky, 1988, 1989; Powell, 2004b; Powell et al., 2005). Descendants of those animals might have interbred with animals descended from ancestors that arrived via natural overwater dispersal and animals introduced more recently, many in association with the burgeoning pet trade. Because of this complex and poorly documented history, whether particular populations of some species were established with human mediation cannot be determined with any certainty. Similarly, house geckos (*Hemidactylus mabouia*) are of African origin (e.g., Kluge, 1969; Vanzolini, 1978). Whether American populations were established as a consequence of natural trans-Atlantic dispersal (see discussion in Mausfeld et al., 2002) or were human-mediated is unknown (e.g., Hedges, 1996). Late Quaternary fossils on Guadeloupe (Pregill et al., 1994) are indicative of a prolonged presence in the region, although Breuil (2002, 2009) noted that only one species of gecko (*Thecadactylus rapicauda*) was known from the region at the time of colonization. However, once established in the Western Hemisphere, populations might have dispersed naturally to Caribbean islands; and such dispersal might have been facilitated by human activities or extant populations might be descendants of ancestors arriving by both means. Herein, we include only peripheral (i.e., Greater Antilles) or recent (Aruba, Bonaire, Curaçao) records, with the implicit assumption that at least some of the Lesser Antillean and Virgin Island populations are natural, although they might frequently be supplemented with individual stowaways.

Economic indicators and the live animal trade

In general, the magnitude of the invasive species problem is proportional to connectivity — the more transport and commerce between two locations, the greater the risk of species being moved (Perry and Vice, 2009, and references therein). This generality has rarely been tested in connection with specific geographic areas (but see Pyšek et al., 2010). To test it for the Caribbean, we compared aggregate arrival data for those species for which such data were available (from Kraus, 2009) to economic indicators for the US and the region for the same period. Although data were not available for the entire time period for which information on introductions is available (1800 onward), we obtained data on the Gross Domestic Product (GDP) during part of this period and the Consumer Price Index (CPI) for the entire period for the mainland US, and the GDP for Puerto Rico (PR), Dominican Republic (DR), and Jamaica (JA) since 1965. Data on US GDP were collected from 1929 onward and obtained from the US Department of Commerce,

Bureau of Economic Analysis (www.bea.gov/national/nipaweb/) and data on US CPI from 1800 onward were obtained from the U.S. Department of Labor, Bureau of Labor Statistics (www.minneapolisfed.org/community_education/teacher/). GDP data for Caribbean economies were obtained from World Bank Reports (http://web.worldbank.org/WBSITE/EXTERNAL/DATASTATISTICS/). Data on the live animal trade during the years 1998-2008 were obtained from the US Fish and Wildlife Service (USFWS) Law Enforcement Management Information System (LEMIS) database. This database only records animals coming into or exported out of the US, and thus represents an underestimate of the total legal traffic in the region, but is the only available source of such information.

Results

Our literature review and ancillary information provided documentation for 364 introductions of 130 species: 25 amphibians (19.2%) and 105 reptiles (80.8%) in the greater Caribbean (appendices 1 and 2). Of those introductions, 256 (70.3%) resulted in populations that were at least temporarily established. Subsequent failures of populations established for at least short periods of time have been recorded in only 29 instances (including six introductions for research purposes). Excluding unidentified species of *Trachemys* in the Bahamas, for which origins are unknown, and species introduced for conservation or research purposes (all of which originated from within the region), 38 species (33.3%) were native to other Caribbean islands and 76 (66.7%) presumably were native to areas outside the region. Most of the latter originated in CA or SA ($n = 32$; 42.1% of those from beyond the greater Caribbean) or NA ($n = 25$; 32.9%), but 19 (25.0%) were from the Eastern Hemisphere. Some species might have been established by individuals from regional captive-breeding programs supplying the international live animal trade, and some Caribbean populations of *Rhinella marina*, *Iguana iguana*, and *Gymnophthalmus underwoodi* might be native, but their exact origins remain unclear. A growing number of introductions (at least 39; 10.7% of all introductions) represent species that originated in the Caribbean or elsewhere, became established outside their native ranges — most notably in Florida — and were then introduced in the region. Nearly all are attributable to three species (*Osteopilus septentrionalis*, *Anolis sagrei*, *Ramphotyphlops braminus*).

Most species have become established on only one or two islands, but 25 species have been introduced to at least three islands or island groups in the region. Although many of the introduced populations are limited to human-dominated habitats, such as urban areas, at least some (e.g., *Rhinella marina*, *Eleutherodactylus johnstonei*, *Iguana iguana*, *Anolis sagrei*, *Boa constrictor*) have successfully invaded natural habitats. Known effects on native species in the region include predation, competition, hybridization, confounding conservation/education programs, and possibly introducing alien disease vectors.

Strays (documented arrivals of one or a few individuals with no evidence of reproduction) represent 24.2% of all documented introductions. These include 8 introductions of amphibians and 80 of reptiles, plus 3 amphibian and 18 reptilian introductions for which the status is unknown and which are presumed to have been strays. Including those would increase the percentage to 29.9% of all introductions.

Although some introduced populations stem from multiple arrivals and the origins of many are unknown, primary pathways for introduction include inadvertent arrivals in cargo and ornamental plants (ca. 100). However, a substantial number are associated with the pet trade (ca. 65). Some of the latter might have been intentional, but most releases were probably accidental. Tortoises (*Chelonoidis carbonaria*) and iguanas (*Iguana iguana*) are widely distributed throughout the region, and many populations probably have mixed origins, with some tracing their ancestry back to individuals that arrived via natural over-water dispersal, intentional introductions by Amerindians and early European colonists, inadvertent releases of pets, or some combination thereof (e.g., Censky, 1988; Powell, 2004b). Complicating matters further are recent intentional inter-island introductions such as that of *C. carbonaria* onto St.-Barthélemy from Saba after World War II (Breuil, 2004) or within the British Virgin Islands (BVI) for conservation purposes (Lazell, 2002, 2005, 2006; Perry and Gerber, 2006).

Although some unintentional introductions occurred more than a century ago, most are more recent. Intentional introductions fall into four broad categories: for food (10 amphibian introductions plus an undetermined percentage of arrivals of *I. iguana*, *C. carbonaria*, and turtles in the family Emydidae), for pest control ($n = 19$, *R. marina* and several instances involving *Eleutherodactylus* spp.), research ($n = 6$), and conservation ($n = 23$). Unlike recent conservation and research-related introductions (all after 1970), intentional introductions for food and biocontrol almost always occurred earlier, many during the 19th century. Rates of new arrivals of both amphibians and reptiles have markedly increased over time (fig. 1). Using only those data for which arrival dates have been documented, arrival rates of the two groups are highly and significantly correlated with each other (Kendall's tau; $n = 13$, $\tau = 0.60$, $P = 0.005$), although the amphibian data represent a smaller number of species than the more taxonomically diverse reptilian data.

Economic activity and herpetological introductions

Rates of arrival for both amphibians and reptiles are correlated with US GDP (Kendall's tau, $n = 8$, amphibians: $\tau = 0.64$, $P = 0.026$; reptiles: $\tau = 0.69$, $P = 0.018$; both amphibians and reptiles: $\tau = 0.69$, $P = 0.018$) and US CPI (fig. 1; $n = 13$, amphibians: $\tau = 0.61$, $P = 0.003$; reptiles: $\tau = 0.62$, $P = 0.003$; both: $\tau = 0.66$, $P = 0.002$), although numbers of introduced species began increasing before either economic indicator. The relationships with regional economic indicators for PR, the DR, and JA were similar, but not statistically significant, attributable both to the smaller sample sizes and slightly more irregular trends in economic

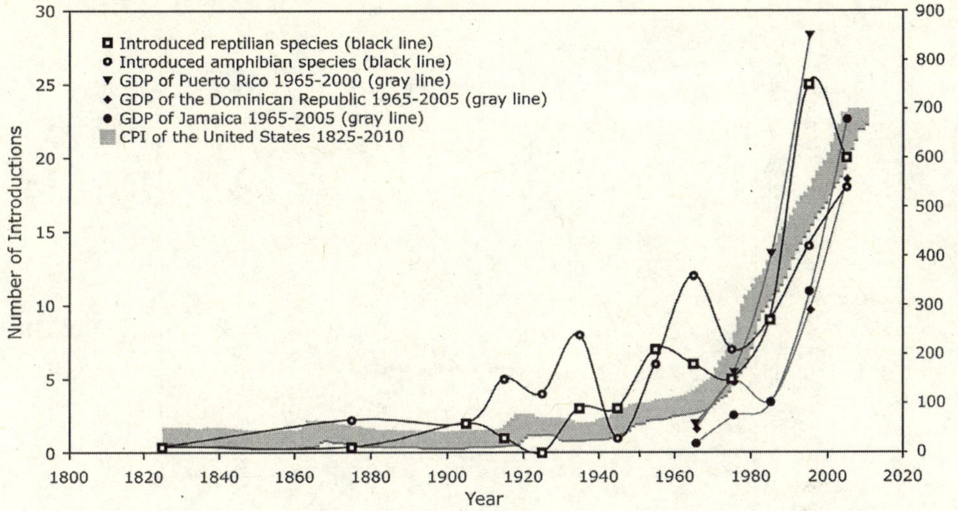

Figure 1. Rates of new amphibian and reptile arrivals (species/decade) for the greater Caribbean from 1825-2005. Note that the first two values for numbers of introductions represent data for fifty-year blocks, the last value is based on pro-rated data for part of the decade, and all of the others are based on data for full ten-year increments. Both amphibian and reptilian arrival rates (black lines) are strongly and significantly correlated with economic activity (broad gray line) in the United States (CPI: consumer price index). Data for the GDP of three Caribbean nations (gray lines) show a similar trend. See text for data sources.

activity. Still, the realities that nearly one-third of the species introduced into the region are from NA and that over 10% of all introductions presumably originated from introduced populations in Florida, in combination with the vast number of amphibians and reptiles transported between the US and the greater Caribbean (see below), suggest that the risk factors for introductions have steadily risen. Consequently, the driving power of US economic activity as a continuously rising source of tourism to the region, a steady destination for regional exports, and a source of imports implicate US and regional economic growth as a major factor directly or indirectly responsible for additional introductions in the future.

Although the Caribbean has never been a large legal export market for amphibians and reptiles for the US, a surprisingly large number of animals are shipped from the US to the Caribbean (appendix 3), the Caribbean to the US (appendix 4), and from the Caribbean to the US and then onwards, including back to the Caribbean (appendix 5). Between 1998 and 2008, approximately 1150 amphibians and 12,650 reptiles were exported each year from the US to the Caribbean. Over that same period, 21 amphibian and 50 reptilian species were exported to the Caribbean, with cumulative numbers of species increasing during that period (fig. 2).

Although some of the species transported from the US into the greater Caribbean (appendix 3) are for conservation purposes (e.g., release of captive-bred *Peltophryne lemur* into PR, where they are native) and others involve trade in species used for food (e.g., *Lithobates catesbeianus* into the DR), a large number are not found in

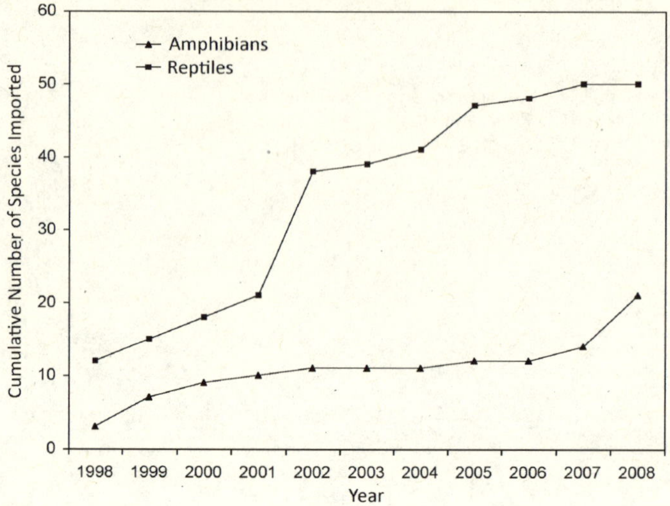

Figure 2. Cumulative number of amphibian and reptilian species exported from the US to the Caribbean between 1998 and 2008. Data are from the USFWS Law Enforcement Management Information System (LEMIS) database.

the region either naturally or as previously introduced populations. Twenty-three species of amphibians (including as many as ten species of salamanders, which are not known to occur naturally on any of the islands) obviously are being shipped to serve the pet/aquarium trade. Although most numbers are relatively modest, 3612 Oriental fire-bellied toads (*Bombina orientalis*) were shipped to the DR in 1999 and 1205 Japanese fire-bellied newts (*Cynops pyrrhogaster*) were sent to the Cayman Islands over a five-year period.

A somewhat similar pattern applied to reptiles exported from the US into the region. A small number are repatriated captive-bred animals, e.g., the *Cyclura nubila* shipped to the Cayman Islands in 1999 are almost certainly *C. lewisi* (then considered a subspecies of *C. nubila*) sent to augment the *in-situ* captive-breeding program. Turtles, which almost certainly represent a combination of animals destined for the pet trade and those destined for food markets, were shipped in the largest numbers, including 12,300 *Pseudemys* sp. sent to the Netherlands Antilles (no indication of whether these are the Leeward or Windward islands of that nation) and 20,066 and 97,910 *Trachemys scripta* shipped to the Bahamas and the DR, respectively. However, 33 transported reptilian species are not currently known to occur on any of the islands, including two cobras (*Naja* sp.) sent to the Bahamas.

Countries within the Caribbean region have exported introduced amphibians and reptiles to the US, as well as countries within the European Union and Asia. Like exports from the US, a few of the species imported from the region into the US (appendix 4) involve conservation programs (e.g., *C. "nubila"* from the Cayman Islands), some are destined for the food/restaurant market (e.g., nearly three million *Lithobates catesbeianus* from the DR), but most almost certainly supply the pet/aquarium trade. Fifteen species are not known to occur in the region, including two cobras (possibly the same animals exported to the Bahamas that same year) and a number of species that are known from the greater Caribbean do not occur on the islands from which shipments into the US originated.

Over half of all 132 records of exports from the US to the region (appendix 3) are to two nations, Barbados with 27 (20.5%) and the Netherlands Antilles with 47 (35.6%). Although a market for pets exists in both nations, active animal vendors are based in those countries, suggesting that many of the exported animals are destined for markets elsewhere. Of 95 records of imports from the region into the US (appendix 4), most come from Barbados (12; 12.6%) and Haiti (26; 27.4%). The former nation, as noted previously, is very active in the pet trade and the latter is a major source of some popular species (e.g., 6720 presumably native *Anolis* spp. and 88,524 *Leiocephalus* spp.) as well as a transshipment center for species from elsewhere within the region (e.g., 56,656 *L. carinatus*, which are not native to Hispaniola) and beyond (e.g., 2623 African *Agama agama*). Of considerable concern are 888 endangered *Osteopilus vastus* shipped from Haiti in a two-year period. These large Hispaniolan endemics are associated with disappearing Hispaniolan gallery forests (Hedges et al., 2004) and have been included among species of special concern in the DR (Powell et al., 2000). The largest number of records of species imported into the US from the region for re-exportation primarily to Europe and Canada (appendix 5) come from Haiti (30 of 49; 61.2%). Although many presumably are native to Hispaniola (e.g., some *Anolis* spp. and *Leiocephalus* spp.), many others (e.g., *Agama agama*, *L. carinatus*) are not. Reinforcing the concept of Haiti, especially, as a transshipment center, several species re-exported from the US are species that are native to that nation.

Taxonomic patterns: Amphibians

Although some urodeles and a variety of frogs are exported to the Caribbean from the US, all amphibians introduced in the region to date have been frogs belonging to six families: Bufonidae, Eleutherodactylidae, Hylidae, Leptodactylidae, Leiuperidae, and Ranidae (until recently, genera in the families Eleutherodactylidae and Leiuperidae were assigned to the family Leptodactylidae). Relatively few genera are represented, most originating from within the region and all from within the Americas. Inadvertent introductions via the nursery trade are the most frequent mechanisms of arrival, although stowaways in cargo are common, as are species arriving via the pet trade and as a consequence of intentional releases for food or biocontrol.

True toads (family Bufonidae). The cane toad (*Rhinella marina*, formerly *Bufo marinus* or *Chaunus marinus*), native to the Neotropics, has been intentionally introduced for biocontol of insect pests in many parts of the world. Although it rarely fulfills that purpose, it feeds voraciously on almost everything else (e.g., Wolcott, 1937; Lynn, 1940; Long, 1974; Breuil, 2002; Meshaka and Powell, 2009), with broad ecological impacts reported from Australia, Florida, and Hawaii (e.g., Esteal, 1981). Wilson et al. (2010) also reported negative effects on native predators, describing mortality in endemic and threatened Jamaican boas (*Epicrates subflavus*) after ingesting cane toads.

The cane toad is widely established in the Caribbean and some populations might be traced to founders that arrived naturally via over-water dispersal (Henderson and Powell, 2009). These toads are ubiquitous on many islands (e.g., Mallery et al., 2007 for St. Vincent). However, populations have failed to become established on islands that provide few opportunities to breed, such as Anguilla (Hodge et al., 2003; Hodge et al., 2011) and Union Island in the Grenadines (J. Daudin, pers. comm.). Also, despite efforts to establish the species on Cuba, it was uncommon by the early 1970s (Schwartz, 1972) and has since failed (Schwartz and Henderson, 1991; Henderson and Powell, 2009). Claims of its past presence in the British Virgin Islands (BVI; MacLean, 1982) have not been confirmed in recent years (G. Perry, unpubl. data; C. Petrovic, pers. comm.). These toads are common commensals, often utilizing human-created habitats such as parks, gardens, and resort grounds (Powell and Henderson, 2008) and exploiting the artificial night-light niche (Perry et al., 2008).

Treefrogs (family Hylidae). Treefrogs are frequently found in the pet trade in North America (NA), but means of dispersal such as stowing away in cargo and arriving with ornamental plants are much more common in the Caribbean. Several species are now found in the region, the most problematic of them being the Cuban treefrog (*Osteopilus septentrionalis*). These frogs readily act as human commensals and have a catholic diet that includes vertebrates (e.g., Meshaka, 2001; Owen, 2005; Powell and Henderson, 2008). Rödder and Weinsheimer (2010) indicated that the entire Caribbean Basin could provide suitable habitat under current climatic conditions. Severe ecological effects are likely, especially when these frogs invade relatively natural areas. For example, locals in the BVI often associate the arrival of *O. septentrionalis* with the ensuing decline and disappearance of native frogs (Owen, 2005). This species now has a broad Caribbean distribution and has also been established elsewhere in the world. The means of arrival are often complex, as single populations might have multiple temporal and geographic origins (e.g., van Buurt, 2007). The population on Anguilla was traced to containers of ornamental plants from Florida, and a small population had been present for several years before generating wide attention after a series of particularly wet years during the late 1990s. At that time, the frogs spread from localized sites (often on resort grounds) to much of the island, where they used various sources of water, including cisterns associated with residences, for breeding (Townsend et al., 2000; Hodge et al., 2003). A similar scenario played out on St.-Barthélemy, where an initial association with resorts was documented by Breuil (2002), Breuil and Ibéné (2008), and Breuil et al. (2009). In the BVI, a Beef Island nursery was a common cause of dispersal, and the owner stated: "These are my children," and refused to take action against the population (J. Owen, pers. comm.). Populations elsewhere have exhibited similar patterns, remaining relatively obscure until propitious weather conditions (often associated with hurricanes) result in a population explosion. Cuban treefrogs were relatively rare on St. Maarten/St.-Martin in the 1980s, but had become almost ubiquitous by the early 1990s (e.g., Powell et al., 1992). Similarly, frogs were infrequently encountered on Antigua until they became a plague during a relatively short period in the late 1990s and early 2000s (Daltry, 2007, 2011; R. Powell, unpubl. data). Spread of this species continues (e.g., Powell, 2006, 2007 on Saba, presumably from St. Maarten; Perry, 2009a on Guana in the BVI, almost certainly from Beef Island). More recently, the species might now be established in the Turks & Caicos Islands (Reynolds and Niemiller, 2010; Reynolds, 2011). In dry years, frogs are less evident (Powell and Henderson, 2008; Hodge et al., 2011). Some populations on Anguilla and in the BVI have shrunk or disappeared as a result of management efforts — primarily blocking access to freshwater sources needed for reproduction — and a regional drought in 2009 (Hodge et al., 2011; G. Perry, unpubl. data).

The exact sources of established populations of *Hyla cinerea* on PR and *H. squirella* in the Bahamas are not known, but the source is likely to have been NA, where both are native. *Pseudacris crucifer* is another NA species. Although reported from Cuba, no extant populations are known to exist. *Scinax ruber* has become established in Martinique, PR, and St. Lucia, but the means of arrival remain largely uncertain. The population of this SA native on St. Lucia appears to have resulted from cargo stowaways (Kraus, 2009). The closely related *S. x-signatus*, also SA in origin, was recently reported on several islands in the Guadeloupean Archipelago (Breuil, 2004) and on Martinique (Breuil, 2011).

Rain frogs (family Eleutherodactylidae). Rainfrogs (genus *Eleutherodactylus*) are among the most commonly introduced amphibians, with the genus and two species listed among the most successful colonizers by Bomford et al. (2009). That success is largely attributable to their frequent association with nursery plants (e.g., Kraus, 2009). At least six different species have been introduced within the Caribbean. *Eleutherodactylus antillensis*, native to PR and the Virgin Islands, is established on St. Croix, US Virgin Islands (USVI; Platenberg and Boulon, 2006) and occurs locally on Necker Island (BVI; Perry and Gerber, 2011). The source appears to have been intentional introductions. *Eleutherodactylus coqui* from PR is established in the DR and the USVI. Initially imported with nursery plants, populations often spread as a consequence of locals intentionally introducing individuals into their yards and gardens. Economic and possibly environmental effects from populations established in Hawaii are considerable (e.g., Kaiser and Burnett, 2006), but similar data from the Caribbean are not available.

Eleutherodactylus johnstonei, originally described from an introduced population on Grenada (Barbour, 1914), is widely distributed in the Lesser Antilles (LA) and also has become established outside the region. Although these frogs do not penetrate high-quality closed-canopy forests in Jamaica (Wilson, 2011), introduced populations often are phenomenally successful. Germano et al. (2003) noted that during a nighttime trip across Grenada, they were out of earshot of calling *E. johnstonei* for only a few seconds in the most densely developed center of St. George's, and Mallery et al. (2007) found calling frogs at every site they sampled on St. Vincent. The nursery trade and stowaways appear to be the primary vectors for dispersal.

Eleutherodactylus lentus, a USVI native, was recently reported from Jost Van Dyke in the BVI (Perry, 2009b). Although the species may have been native there, it most likely is a recent introduction via construction materials. Calling individuals located over multiple years suggest that this population is well established. *Eleutherodactylus martinicensis* from Antigua, Guadeloupe, Dominica, and Martinique was established on St.-Barthélemy as a result of the nursery trade (Kaiser, 1992) and on St. Maarten/St.-Martin, either via the nursery trade or as a stowaway in other cargo (Breuil, 2002). *Eleutherodactylus planirostris* from Cuba and the Bahamas is also broadly established, both within the Caribbean and beyond, including new populations in the Turks & Caicos (Reynolds and Niemiller, 2010; Reynolds, 2011). Although the nursery trade is involved in many instances, unintentional arrival via cargo also has been documented (Kraus, 2009). *Eleutherodactylus cochranae* from PR was introduced for research purposes onto Isla Palominitos, but the introduction did not result in an established population (Levins and Heatwole, 1973). However, *E. schwartzi* was intentionally translocated for conservation purposes within the BVI from Great Dog to Little Thatch Island, where it is now established (Lazell, 2005).

True frogs (family Ranidae). These frogs in the genus *Lithobates* (formerly assigned to the genus *Rana*) are associated with permanent bodies of (often flowing) water and would appear to be poor candidates for introduction. However, five species have been reported from the Caribbean, all of NA origin. The most widely distributed, and also potentially the most damaging, is the American bullfrog (*L. catesbeianus*), which was listed as the fourth most successful colonizing species by Bomford et al. (2009). Populations in the region almost certainly were intentionally introduced for food, although the pet trade has been implicated as a source of some populations in Canada and Europe (Kraus, 2009). The species is a major export out of the Caribbean, presumably also for the food industry. More aquatic than most other anurans in the region, the densest populations often are associated with artificial habitats such as drainage ditches, water hazards on golf courses, and reservoirs (Powell and Henderson, 2008), although they have successfully exploited natural bodies of water as well. Reaching a very large size, this species is capable of ingesting bats (Vogel, 1965) or birds (López-Flores et al., 2003), although much of the diet of West Indian populations is comprised of invertebrates (Pérez, 1951; Mahon and Aiken, 1977; Sampedro Marín et al., 1985, 2003; Montañez Huguez et al., 1996). At least two instances of bullfrogs consuming native West Indian frogs have been documented, *Leptodactylus albilabris* in Puerto Rico (Thomas and Joglar, 1996) and *Osteopilus dominicensis* on Hispaniola (Neils and Bugbee, 2007). Schloegel et al. (2009) implicated bullfrogs as vectors for *Batrachochytrium* *dendrobatidis* and ranavirus introduced into the US and elsewhere. Considering the massive numbers of frogs moving from the Caribbean back and forth to the US and other parts of the world, they might very well be the sources of many chytrid infections that are being documented with increasing frequency in the region (e.g., Henderson and Powell, 2009).

The status of *L. clamitans* in the Bahamas remains uncertain (Knapp et al., 2011), but this species also attains considerable size and has the potential to cause ecological damage. *Lithobates grylio* is established in the Bahamas and Puerto Rico, and almost certainly was introduced intentionally as a human food source. *Lithobates sphenocephalus* apparently is established in the Bahamas, but *L. pipiens* failed to become established on St. Croix. The modes of introduction for these populations are unknown, but some were almost certainly intentional.

Narrow-mouthed frogs (family Microhylidae). Frogs in the genus *Gastrophryne* are secretive NA species with which most people are unfamiliar because they are primarily fossorial. One species, *G. carolinensis*, nonetheless succeeded in establishing itself in the Bahamas and on Grand Cayman Island, having arrived with ornamental plants (Seidel and Franz, 1994).

Neotropical frogs (family Leptodactylidae). Native to the Caribbean, the very robust "Mountain Chicken" (*Leptodactylus fallax*) has been introduced on Grenada, Jamaica, Martinique, and Puerto Rico, presumably intentionally as a delicacy (Kraus, 2009). All attempts ultimately failed, although the introduction to Martinique might date to Amerindians (Breuil and Ibéné, 2008; Breuil et al., 2009). Ironically, this species is rapidly declining in its native range (e.g., Garcia et al., 2007). Recent work (Yanek et al., 2006; Camargo et al., 2009) suggested that *L. validus*, long believed to be native to St. Vincent and Grenada, was in fact introduced into the LA with early human arrivals.

South American foam-nesting frogs (family Leiuperidae). *Pleurodema brachyops* is a SA species that has been on Aruba for a long time (its native Caquetío name, Dori Maco, is not used in Venezuela; van Buurt, 2005). The species might be native there, although introduction by Amerindians cannot be ruled out. Populations on Curaçao arrived largely with sand dug from the bottoms of water reservoirs on Aruba, which was used as grit to sandblast steam boilers (van Buurt, 2001, 2005). Those on Bonaire originated on Curaçao, arriving as tadpoles brought back from Curaçao and intentionally released in a small reservoir (van Buurt, 2001, 2005).

Taxonomic patterns: Reptiles

A variety of reptilian taxa has arrived in various Caribbean locations, and disconcertingly large numbers of those have become established. A large proportion of these species is of regional origin, although some originated in the Eastern Hemisphere. The two primary paths of arrival appear to be stowaways in cargo and, more recently, the pet trade, although other sources have been reported.

Crocodilians (families Alligatoridae and Crocodilidae). Such large and obvious animals might seem unlikely to be invasive, since they are not likely to stow away unnoticed. Several species are found in the pet trade, however, and this is the likely source of most records of *Caiman crocodilus*, although the population on Isla de la Juventud, Cuba, was intentionally introduced as a potential source of hides and meat (Soberon et al., 1996; Kraus, 2009). Other observations of non-native crocodilians in the region are of strays (e.g., "an undetermined caiman from Guiana" on Martinique; Breuil, 2009), none of which have become established. In general, such arrivals remain uncommon, both in terms of numbers and geographic scope.

Tortoises (family Testudinidae). South American tortoises in the genus *Chelonoidis* (formerly assigned to the genus *Geochelone*) tend to be large and are introduced primarily via the pet trade or as ornamentals, although their willingness to consume fecal matter renders them useful for cleaning latrines or chicken pens (e.g., Grant, 1937a; Pinchon, 1967). Daudin and de Silva (2007, 2011) indicated that locals in the Grenadines scorn them as food for that very reason. The status of *C. carbonaria* populations on many islands remains unclear (e.g., Censky, 1988; Hodge et al., 2003; Powell et al., 2005; Powell and Henderson, 2005; Fields and Horrocks, 2009), with the ancestors of some likely arriving via natural over-water dispersal, whereas those of others might have been introduced by Amerindians or early colonial Europeans (perhaps for food), and others being more recently moved for conservation (under the assumption that they are declining natives; Lazell, 2002, 2005; Perry and Gerber, 2006) or as pets and for ornamental value (e.g., Breuil, 2002; Powell et al., 2005; Lorvelec et al., 2007). Individuals from Barbados are exported regularly to supply the pet trade (Fields and Horrocks, 2009). The closely related *C. denticulata*, originally from SA, is introduced on Guadeloupe (Pritchard and Trebbau, 1984; Breuil, 2002), although only escaped individuals are known (i.e., no feral population exists). *Centrochelys sulcata* (also previously assigned to the genus *Geochelone*), from northern Africa, is known as a stray on Martinique (Breuil, 2009). The latter has also been brought in as an ornamental on several islands in the BVI (G. Perry, unpubl. data). Early 19th-century attempts to establish two species of *Kinixys* (*K. erosa* and *K. homeana*) on Guadeloupe failed (Breuil, 2002, 2003).

Pond turtles (family Emydidae). Pond turtles of several species are common in the pet trade, which is the primary vector for their worldwide spread, although some populations are exploited for food (e.g., Powell, 2003). *Pseudemys nelsoni* from NA, presumably released pets, were removed from one location in the BVI before they could breed (Perry and Gerber, 2006). *Graptemys pseudogeographica*, also from NA, is known as a stray on Martinique (Breuil, 2009). Much more broadly distributed, however, is *Trachemys scripta*, another NA species. This is one of the most common species in the pet trade and also is marketed for food, to such an extent that multiple arrivals at any given location are not unlikely. Many Caribbean populations, such as those in the Turks & Caicos (Reynolds and Niemiller, 2010;

Reynolds, 2011) and BVI (Perry and Gerber, 2006) appear localized in human-made structures and are unlikely to have extensive impacts on native species. Even where abundant (e.g., St. Maarten/St.-Martin; Powell et al., 2005), severe ecological effects are unlikely where no native congeners occur. Where native sliders occur, however, such as the Bahamas, Cuba, Hispaniola, and Jamaica, hybridization and competition are both of concern (Powell et al., 2000; Powell and Incháustegui, 2009, 2011), as is the possibility that efforts to conserve the native species will be confounded by confusion with invasives that should be controlled. Seidel and Ernst (2006) noted that the extent to which the introduction of *T. scripta* "promotes extirpation or extinction by genetic 'swamping' is an overlooked topic in turtle conservation which deserves attention."

Four other members of the genus *Trachemys* (*T. decorata*, *T. decussata*, *T. stejnegeri*, and *T. terrapen*), all from within the West Indies, have become established at some locations outside their native ranges. Although some introductions have failed (Kraus, 2009), interbreeding "swarms" of hybrids are suggestive of multiple invasions on New Providence, Andros, and Great Exuma islands in the Bahamas (Schwartz and Henderson, 1991; Franz et al., 1993; Lee, 2004, 2005). The status of *Terrapene carolina* in the Bahamas (Lee, 2004, 2005) and on Martinique (Breuil, 2009) is unknown. This terrestrial member of the family originated in NA and Caribbean populations almost certainly are pet-trade related.

Afro-American side-necked turtles (family Pelomedusidae). *Pelusios castaneus* is African in origin and not uncommon in the pet trade. Lescure (1979) indicated that the exact source of the population established on Guadeloupe (e.g., Lescure, 1979, 1983) remains unclear, but Breuil (2003) indicated that it was introduced intentionally by L'Herminier in the early 19th century.

Austro-South American side-necked turtles (family Chelidae). No documented explanation exists for the single *Phrynops geoffroanus* found on Anguilla (Hodge et al., 2011).

Geckos (family Gekkonidae). Many geckos are common human commensals that have become widely distributed around the globe. The genus *Hemidactylus* and four species (three of which are known from our region) are included among the most successful colonizers (Bomford et al., 2009). Until recently (Weiss and Hedges, 2007), *H. haitianus* was considered a West Indian endemic (Powell et al., 1996). Now known to be conspecific with African populations of *H. angulatus*, the likelihood of a relatively recent human-mediated introduction into the Greater Antilles (possibly with the slave trade) is high. The history and movement of populations within the Greater Antilles is unknown. *Hemidactylus garnotii* is established on several Bahamian islands, having arrived as a stowaway. The most widely distributed "house gecko" within the region, *H. mabouia*, is found on many islands, where it is essentially ubiquitous on buildings and walls (e.g., Howard et al., 2001). Origins are uncertain (e.g., Kluge, 1969; Powell et al., 1998); although some insular populations might have arrived via natural over-water dispersal from SA (or even Africa), others probably arrived with humans, and some populations are likely mixtures of both. Breuil (2009) recommended studies using molecular markers to identify the origins of insular populations in the region. On Curaçao and Bonaire, these invasives are displacing the native *Phyllodactylus martini*, which is no longer edificarian and is only infrequently encountered in the bush, often in the wood of old candelabra cacti (van Buurt, unpubl. data). House geckos continue to invade additional islands, such as the Turks & Caicos (Reynolds and Niemiller, 2010; Reynolds, 2011). Another member of this genus, the Mediterranean *H. turcicus*, occurs locally on Cuba. Whether that population originated from Europe or from US populations is unknown. Most of the Caribbean populations of all of these species are probably derived from other non-native populations in the region or adjacent mainland areas. *Hemidactylus frenatus*, native to the Eastern Hemisphere, but widely established in the Americas, has recently been found on Hispaniola (Scantlebury et al., 2010) and at the U.S. Naval Base at Guantanamo Bay, Cuba (S. Campbell-Staton, pers. comm.). *Hemidactylus palaichthus* is a Neotropical endemic (Kluge, 1969), with populations in northeastern SA, adjacent continental islands, and on the Maria Islands off St. Lucia (Powell, 1990c). Originally thought to be derived from *H. brookii haitianus* (= *H. angulatus*; see above), its current status is uncertain. Whether the Maria Island population is of natural or anthropogenic origin is unknown.

Intentionally introduced on Martinique (Henderson et al., 1993) and now also known from Guadeloupe (Breuil, 2009), *Gekko gecko* is the only member of this Asian genus to invade the Caribbean.

Dwarf geckos (family Sphaerodactylidae). The genus *Gonatodes* contains mostly diurnal species widely distributed throughout the Neotropics. The founders of some Greater Antillean populations of *G. albogularis* almost certainly reached the islands by natural means, although Crombie (1999) suggested that: "Its distribution around major centers of human habitation in both Jamaica and Hispaniola smacks of an introduction." Populations on Aruba and Curaçao (Wagenaar Hummelinck, 1940) originated in SA, although they probably are no longer extant (Lundberg, 2003; G. van Buurt, unpubl. data). The population on Grand Cayman Island probably came from Cuba (Williams, 1964; Seidel and Franz, 1994). Whether this occurred naturally or with human mediation is unknown. *Gonatodes antillensis*, a species with presumably native populations on Bonaire and Curaçao, might have spread to Aruba via human actions (Odum, 1992), but has not been found there recently (Wagenaar Hummelinck, 1940; van Buurt, 2001, 2005; Lundberg, 2003). *Gonatodes vittatus* is very common on Aruba (where it might be native), but is rarely found on Curaçao. The origin of a single individual on Dominica was probably Venezuela (Malhotra et al., 2007, 2011).

Sphaerodactylus geckos are small, frequently diurnal, often commensal lizards that have speciated widely in the region. Many species occur naturally in the islands, and a few have become invasive, spreading primarily as stowaways in cargo (Kraus, 2009). These include *S. argus*, *S. copei*, *S. mariguanae*, and *S. microlepis*. Evans (1989) suggested that *S. fantasticus* was introduced on Dominica, but evidence (Jones, 1999; Malhotra et al., 2007, 2011; Thorpe et al., 2008) indicates that *S. fantasticus* is a relatively recent (but probably pre-human) colonizer on Dominica. The population there appears very similar to those on western Basse Terre, Guadeloupe (e.g., Daniells et al., 2010).

Iguanas (family Iguanidae). Although most West Indian populations of *Cyclura* are Endangered or even Critically Endangered (IUCN, 2010), they are sometimes associated with the pet trade. The zoo trade was responsible for a successful introduction of Cuban *C. nubila* on Isla Magueyes off Puerto Rico (e.g., Schwartz and Carey, 1977; Kraus, 2009), but the increased abundance on the main island (M.J. Rivera Rodríguez, pers. comm.) is probably attributable to active dispersal from Isla Magueyes. The same species presumably was introduced to Grand Cayman Island as a food source (Grant, 1940), possibly a response to declining numbers of endemic *C. lewisi*. Tourists apparently are responsible for the relocation of *C. cychlura inornata* from Bahamian cays to nearby, previously unoccupied cays (Hines and Iverson, 2006a, 2006b). Unfortunately, the substrate on many of the latter precludes nesting, rendering the "relocated" populations biologically dead. Other relocations in the Bahamas, Turks & Caicos, and BVI were motivated by conservation concerns and an effort to render remaining populations less vulnerable to stochastic events.

Like some tortoises, West Indian *Iguana iguana* populations include those founded by ancestors that arrived naturally (e.g., St. Lucia, Saba), some of which might now be distinct at the species level (Malone and Davis, 2004; Powell, 2004b). Other founders were transported by Amerindians or early colonists, have arrived recently, or represent mixtures of the above (Powell, 2004b; Henderson and Powell, 2009). Although some early introductions presumably were for food (e.g., Grant, 1937a), the pet trade is the primary culprit responsible for many of the more recent introductions (Powell, 2004b). These animals pose a threat to endemic Lesser Antillean populations of *I. delicatissima*, with which they hybridize (e.g., Breuil and Sastre, 1994; Day and Thorpe, 1996; Breuil, 2002; Breuil et al., 2007, 2010). The presence of *I. iguana* on Grand Cayman confounds efforts to conserve endemic *Cyclura lewisi*, as some residents and many guest workers do not distinguish one kind of iguana from another (Henderson and Powell, 2009). Economic impacts include the interruption of air travel by individuals basking on runways at the international airport in San Juan, Puerto Rico (Engeman et al., 2005). A few recent introductions have been intentional, and, in at least one case in the Virgin Islands, an introduction was actively facilitated by a local wildlife management agency (Perry and Platenberg, 2007). The report of an *I. delicatissima* introduced to Puerto Rico (Lever, 2003) remains unsubstantiated. A population, however, was introduced from Îlet Chancel to Îlet à Ramiers (Martinique) for conservation purposes (Breuil, 2009). The record of *Ctenosaura similis* in the Bahamas (Knapp et al., 2011) is almost certainly related to the pet trade.

Curly-tailed lizards (family Leiocephalidae). *Leiocephalus carinatus* has been introduced experimentally onto small cays with *Anolis sagrei* to test effects of a predator (Schoener et al., 2005 and references therein).

Anoles (family Polychrotidae). Anoles are highly diverse (Losos, 2009), quite adaptable, and often function as human commensals. Many species in the region exploit buildings, ornamental plants, and the night-light niche (e.g., Henderson and Powell, 2001, 2009; Perry et al., 2008; Powell and Henderson, 2008). Some are colorful and available in the pet trade (e.g., Kraus, 2009), but nearly all introductions within our region were inadvertent and attributable to stowaways in cargo such as building materials and ornamental plants.

Anolis cristatellus is native to the Puerto Rico Bank and was the only anole that made the list of most successful colonizing species (Bomford et al., 2009). A population became established in the DR in the early 20th century (Powell and Henderson, 2008 and references therein). It quickly displaced its native ecological counterpart (*A. cybotes*) from the most intensely altered habitats in and around the city of La Romana. These anoles have more recently been introduced into Dominica (Malhotra et al., 2007, 2011), where they are expanding their range and displacing endemic populations of *A. oculatus* along the dry leeward coast, and to St.-Martin (Breuil et al., 2010). Cuban *A. porcatus* became established in Santo Domingo (DR) in the mid-20th century (Powell and Henderson, 2008 and references therein) and, much like *A. cristatellus* in La Romana, has displaced its endemic ecological equivalent (in this instance, *A. chlorocyanus*) from much of the urban area. *Anolis porcatus* also has been reported from Aruba, to which it probably was introduced with a shipment of palm trees from Cuba (Odum and van Buurt, 2009).

Perhaps the most frequently relocated West Indian member of the genus is *A. sagrei*, which is native to the Bahamas, Cuba, and presumably Little Cayman in the lesser Cayman Islands. This species is established in Jamaica, where its presence was documented as early as the mid-19th century (Gosse, 1850). These aggressive lizards can affect other anoles negatively (e.g., Brown and Echternacht, 1991), and have displaced endemic *A. carolinensis* from much of peninsular Florida (Lever, 2003 and references therein). Nothing comparable appears to be occurring on Grenada (Greene et al., 2002) or St. Vincent (Treglia et al., 2008), where populations have become established with building materials, but so far appear to be restricted to only the most intensely altered habitats on those islands. Whether such constraints will continue to constrain expansion in the future or whether they will apply to recently reported populations on Barbados (Fields and Horrocks, 2009), St. Maarten (Fläschendräger, 2010), and Canouan in the Grenadines (M. de Silva, pers. comm.) is unknown. *Anolis sagrei* is comparable in size to the native species there and the potential for competition and possible displacement exists. A population on Aruba might be extirpated (G. van Buurt, unpubl. data).

Populations of *A. carolinensis*, a NA native, have become established inside and outside of the Caribbean. Although the pet trade has been implicated in many instances (Kraus, 2009), the West Indian introductions all appear to be consequences of arrival with nursery plants (e.g., Eaton et al., 2001; Powell, 2002; Hodge et al., 2003). A number of insular populations initially identified as *A. carolinensis* now are assigned to other species of anoles (Henderson and Powell, 2009).

Anolis extremus from Barbados and *A. wattsi* from Antigua are both established on St. Lucia, where they interact with each other and with endemic *A. luciae* (Lazell, 1972; Gorman, 1976; Henderson and Powell, 2009). Other regional anoles found outside their native ranges include strays (*A. equestris*, *A. garmani*, *A. leachii*) or localized populations not far from their points of origin (*A. distichus*, *A. lineatus*, *A. maynardii*).

The introduction of *A. bimaculatus* in St. Maarten (Powell et al., 1992) appears to be one of the few documented colonization failures in the region (Powell et al., 2005). Researchers intentionally introduced Puerto Rican *A. pulchellus* and *A. stratulus* into Isla Palominitos (Levins and Heatwole, 1973), which is essentially adjacent to both species' native range. Other researchers introduced *A. pogus* from the Anguilla Bank onto Anguillita (Roughgarden et al., 1984). All of those introductions eventually failed.

Ground lizards (family Teiidae). Lizards in the genus *Ameiva* are common on many Caribbean islands. Some species become habituated to human presence and many can be found in urban settings (Henderson and Powell, 2001; Powell and Henderson, 2008). *Ameiva exsul* has become established on St. Croix in the USVI, where it is a source of concern for the critically endangered native congener, *A. polops* (Platenberg and Boulon, 2006). Although St. Croix is part of the USVI, it is not part of the Puerto Rico Bank, which *A. exsul* inhabits. The species can swim and has been seen on cargo barges (Perry et al., 2006), providing a possible transport mechanism. *Ameiva ameiva*, which occurs naturally on the Grenada and St. Vincent island banks, has been documented on Barbados (Fields and Horrocks, 2009), presumably, however, originating from Trinidad. Wagenaar Hummelinck (1940) suggested that *A. bifrontata* was introduced in Aruba, but van Buurt (2001, 2005) indicated that the population might be native, since it was recorded by Cope as early as 1885.

Cnemidophorus lemniscatus, broadly distributed in the Neotropics, has expanded its range onto Aruba by stowing away with cargo (Schall, 1973; van Buurt, 2001, 2005). In contrast, *C. vanzoi* was intentionally introduced to Praslin Island from nearby natural populations, for investigative and conservation purposes (Dickinson and Fa, 2000). The population appears to have successfully colonized its new habitat. *Tupinambis teguixin*, from the Neotropical mainland, has been recorded on Isla de San Andres (Rueda-Almonacid, 1999), but the fate of that introduction is unknown.

Worm lizards (family Gymnophthalmidae). Gymnophthalmids, most occurring in CA or SA, usually are small and many are associated with leaf-litter or live underground (Avila-Pires, 1995). *Gymnophthalmus pleii* is a Lesser Antillean endemic and *G. underwoodi*, which occurs on a number of Lesser Antillean islands, might have reached many of them via natural over-water dispersal (Powell, 2011). However, at least some populations, certainly those in the central and northern LA, are introduced (Powell, 2011), and some might be competing with or even displacing native populations of *G. pleii* on Martinique (Breuil, 2009) or Dominica (Turk et al., 2010). *Gymnophthalmus underwoodi* was recently found for the first time in the Greater Antilles (Hispaniola; Scantlebury et al., 2010).

This species is parthenogenetic (e.g., Cole, 1975; Hardy et al., 1989), which facilitates colonization because single individuals can establish new populations (e.g., Schwartz and Henderson, 1991; Hodge et al., 2003; Powell et al., 2005). *Tretioscincus bifasciatus*, another SA species, has been recorded from Isla de Providencia (Scott and Ayala, 1984; Ayala, 1986; Schwartz and Henderson, 1991; Rueda-Almonacid, 1999), but the status of that population is unknown.

Alligator lizards (family Anguidae). The only anguid known to have been introduced in the Caribbean, *Ophisaurus ventralis* from NA, was reported from Grand Cayman (Seidel and Franz, 1994), but has not been seen in many years and may be extirpated (A.C. Echternacht, unpubl. data).

Helmeted lizards (family Corytophanidae). *Basiliscus* sp. on New Providence (Bahamas; Knapp et al., 2011) probably represents a stray introduced via the pet trade.

Monitor lizards (family Varanidae). A single *Varanus exanthematicus* was occasionally spotted on Providenciales (Turks & Caicos) until 2004 and was known to be a released pet (Reynolds, 2011).

Amphisaenians (family Amphisbaenidae). Two records of *Amphisbaena fuliginosa* from SA, presumably strays, are known from St. Lucia and Grenada (Murphy et al., 2010).

Blindsnakes (family Typhlopidae). Usually small and spending most of their lives underground, blindsnakes are unfamiliar to the general public and practically unheard of in the pet trade. However, they easily stow away in planters and often are spread by the ornamental plant trade. Originally from Asia, the flowerpot snake (*Ramphotyphlops braminus*) might be the most widely distributed snake in the world. The family Typhlopidae, genus *Ramphotyphlops*, and species *R. braminus* top the respective lists of most successful colonizing taxa (Bomford et al., 2009). Since the first report of the species on Anguilla (Censky and Hodge, 1997), it has been documented widely in the Caribbean, including recent reports from Aruba (van Buurt, 2006, 2011), St. Christopher (Orchard, 2010a), Barbados (Fields and Horrocks, 2009, 2011), Guadeloupe (Breuil and Ibéné, 2008; Breuil, 2009), Mustique (M. de Silva, in litt., 2009), the Turks & Caicos (Reynolds and Niemiller, 2010; Reynolds, 2011), Curaçao (Wallach, 2008), and St. Eustatius (Powell, 2011). A parthenogenetic species, it appears to be ideally pre-adapted to dispersal by humans (e.g., McKeown, 1996). West Indian populations are almost certainly derived from the introduced population in Florida.

Threadsnakes (family Leptotyphlopidae). *Epictia* (formerly *Leptotyphlops*) *albifrons*, from the SA mainland, is known from Bonaire (van Buurt, 2001, 2005, 2006, 2011).

Boas (family Boidae). Boas are common in the pet trade, which is the primary method of arrival for these species in the Caribbean and elsewhere (Kraus, 2009). Interestingly, Bomford et al. (2009) rated the family Boidae as the least successful colonizing family of reptiles or amphibians. Most records of *Boa constrictor*, presumably from the SA or CA mainland, are of strays that have failed to found populations. The species, however, is breeding in Puerto Rico (Krysko and King, 2010; USFWS, 2010; M.J. Rivera Rodríguez and A.J. Sánchez Muñoz, pers. comm.; R. Reed and G.H. Rodda, pers. comm.) and is successfully established in Aruba (Quick et al., 2005; van Buurt, 2001, 2005, 2006, 2011), where it has become a serious threat to local birds and other animals. The stray found on Terre de Bas (îles de la Petite Terre, Guadeloupe; Barré et al., 1997) might pertain to *Boa nebulosa* (Lorvelec et al., 2011), which is endemic to Dominica. *Epicrates cenchria*, also from SA, is known as a stray on St. Maarten (Powell et al., 2005) and Martinique (Breuil, 2009), and *Eunectes notaeus* has been reported in Puerto Rico (USFWS, 2010; R. Reed and G.H. Rodda, pers. comm.).

Pythons (family Pythonidae). Pythons, like boas, are frequently encountered in the live animal trade. Six records, *Morelia amethistina* on Guadeloupe (Breuil and Ibéné, 2008), *Python curtus* and *P. regius*, both on St. Maarten (Powell et al., 2005), and *P. regius* on Martinique (Breuil, 2009), St.-Barthélemy (Breuil et al., 2010), and Puerto Rico (USFWS, 2010; A.J. Sánchez Muñoz, pers. comm.), document strays. Reports of populations of *P. bivittatus*, *P. reticulatus*, and *P. sebae* in Puerto Rico (ISSG, 2010; USFWS, 2010), and *P. bivittatus* on St.-Barthélemy (Breuil et al., 2010) presumably also pertain to strays. Establishment of any of these species would be worrisome, as it has been in Florida (e.g., Snow et al., 2007; Reed et al., 2010).

Common snakes (family Colubridae). The pet and nursery trades are the primary vectors for arrival of *Pantherophis guttatus* (formerly *Elaphe guttata*) from NA. For example, animals on Little St. James (USVI) arrived in ornamental plants (Perry and Platenberg, 2007), but the pet trade is implicated in most other introductions in the region. The ongoing spread of this very efficient predator is a source of increasing concern, as some populations are showing signs of reproduction (Tolson and Henderson, 2011; Virgin Islands, R. Platenberg, unpubl. data; C. Petrovic, pers. comm.). The status of this species in the Cayman Islands remains unclear (Franz et al., 1987; A.C. Echternacht, unpubl. data). Other records to date document only strays. Another member of the genus, *P. alleghaniensis* (formerly *Elaphe obsoleta*), apparently is breeding in the Bahamas following arrival via the nursery trade (Buckner and Franz, 1994d; Knapp et al., 2011). Similarly, a stray *Opheodrys aestivus* has been documented in the Bahamas (Knapp et al., 2011) and a stray *Leptophis* sp. arrived as a stowaway on Curaçao (van Buurt, 2001, 2005). *Tantilla melanocephala*, probably of SA origin, has been found on four islands on the Grenada Bank (Henderson and Powell, 2006; Berg et al., 2009; D. Scantlebury and J. Boone, pers. comm.). Its long-term prospects remain unclear. Underwood et al. (1999) reported the presence of *Mastigodryas bruesi*, which occurs naturally on the Grenada and St. Vincent banks, on Barbados.

American rear-fanged snakes (family Dipsadidae). *Diadophis punctatus*, originally from NA, arrived via the nursery trade on Grand Cayman (Seidel and Franz 1994; A.C. Echternacht, unpubl. data) and Curaçao (van Buurt, 2001, 2005). Neither appears to have established a population. Other strays on Curaçao include *Imantodes* sp. and *Leptodeira* sp. (van Buurt, 2001, 2005). Similarly, four strays from within the region, *Alsophis rufiventris* from Saba or St. Eustatius on St. Maarten, *Borikenophis* (formerly *Alsophis*) *portoricensis* from the Puerto Rico Bank on St. Croix, a main-island subspecies of *B. portoricensis* on Little St. James (BVI), and *Hypsirhynchus* (formerly *Antillophis*) *parvifrons* from Hispaniola on Little Inagua Island (Bahamas), failed to establish viable populations; however, *B. portoricensis* has recolonized the eastern end of St. Thomas (USVI) and appears to be expanding westward (Platenberg and Boulon, 2011).

Water snakes (family Natricidae). *Thamnophis cyrtopsis* on Curaçao (van Buurt, 2001, 2005), *T. sirtalis* on St.-Barthélemy (Breuil et al., 2010), and *Storeria dekayi* in the Bahamas (Buckner and Franz, 1998a, 1998b; Lee, 2004, 2005), all from NA, arrived in nursery plants. Only *S. dekayi* seems to have become established. Two stray *Natrix natrix*, from Europe, have been recorded on Martinique (Breuil, 2009).

Coral snakes (family Elapidae). A single individual *Micrurus fulvius*, native to the southeastern US, arrived in the soil of a potted palm on Curaçao (van Buurt, 2001, 2005).

Discussion

The number of introductions and the consequent number of established populations in the Caribbean is alarming, even when we consider our uncertainty regarding the origins of some insular populations (particularly some of those of *Eleutherodactylus johnstonei*, *Rhinella marina*, *Gymnophthalmus underwoodi*, *Hemidactylus* spp., *Iguana iguana*, *Chelonoidis carbonaria*), some of which were almost certainly natural, but others undoubtedly were human-mediated or some combination of the two. Several additional reports arrived as we were working on this review, and the trends shown by both amphibians and reptiles (fig. 1) suggest that the rate of arrivals will continue to increase with time. For example, *Scinax* cf. *x-signatus* has been found on five new islands (Grande-Terre, Basse-Terre, Désirade, Marie-Galante, Martinique) in the past eight years (Breuil and Ibéne, 2008; Breuil, 2011). In some instances, these frogs are phenomenally abundant. Multiple invasions of the same islands are almost certainly responsible, with wooden houses prefabricated in Brazil and French Guiana apparently serving as the means of introduction.

Given the extent of negative ecological and economic effects documented in the Caribbean and elsewhere, invasive populations of herpetofauna have become a serious conservation issue. Additional deleterious effects probably go unnoticed or unreported, especially when smaller, less obvious species are introduced and their impact is primarily on smaller invertebrates, which are rarely monitored and the impact on which cannot, at this time, be evaluated. The magnitude of existing problems is almost certainly greater than currently realized, and can only get worse. An integrated policy response is clearly necessary to address what is a regional issue.

Amerindians arrived in the Caribbean islands about 6000 years ago (Wilson, 2001) and Europeans about 500 years ago. The impact of the latter has been felt in the region longer than elsewhere in the Western Hemisphere, and Fosberg (1983) observed that: "The impact of European man on islands made the changes due to aboriginal man seem minor by comparison." Only 5-10% of the West Indian herpetofauna has benefited from human activities (Henderson and Powell, 2001). One of the most substantive and frequently deleterious effects has been the increasing number of introductions of plants and animals to islands where they are not native. Although not covered here, many of those introductions, especially of mammalian herbivores (e.g., goats and cattle) and predators (e.g., mongooses, raccoons, oppossums, dogs, cats, and monkeys, the latter on Grenada and Barbados), have had varying degrees of deleterious effects on the regional herpetofauna.

The characteristics of amphibian and reptilian species introduced in the region correspond very closely to those outlined for taxa associated with Caribbean urban areas by Powell and Henderson (2008): they (1) are ecologically versatile and capable of tolerating a broad range of sometimes rapidly and dramatically changing conditions; (2) exhibit edificarian tendencies within and outside of urban areas (e.g., gekkonids, some sphaerodactyls, many anoles); (3) tend to be edge species or, at least, species that are euryoecious, not habitat specialists; (4) are primarily invertebrate predators; (5) are heliotherms if diurnal (e.g., edge-inhabiting anoles); and (6) often are scansorial (e.g., treefrogs, geckos, anoles). These features in turn correlate nicely with those shared by anoles identified by Williams (1969) as successful colonizers, and with the observation that invasive species in general tend to be generalists (Dukes and Mooney, 1999). Our data also support the generalizations that good invaders tend to be small and capable of rapid reproduction (Kolar and Lodge, 2001), have a past record of being invasive elsewhere (Kolar and Lodge, 2001; Marchetti et al., 2004), are highly tolerant of humans (Perry et al., 2008), are related to other documented invaders (Richardson and Pyšek, 2006), and are native to areas with comparable climates (Bomford et al., 2009) and near possible introduction sites (Marchetti et al., 2004). In contrast, our data do not support the view that taxa that are more distantly related to the native biota are more likely to be invasive (Strauss et al., 2006).

The genera *Eleutherodactylus*, *Hemidactylus*, and *Anolis* comprise a large fraction of the species tabulated in this review. All are relatively small, capable of high reproductive output, often associated with humans and habitats modified by human activities, are naturally or secondarily found within the region, and have become invasive at multiple locations. Specifically, the species that have successfully colonized the most islands either follow that pattern or have been intentionally introduced for perceived economic benefits. Cane toads (*Rhinella marina*; 25 islands or island groups, although some populations might have been established by natural over-water dispersal) were introduced widely for biocontrol, and bullfrogs (*Lithobates catesbeianus*; all of the Greater Antilles) were introduced for food. The latter are still exported from the region in large numbers, providing an economic incentive for further spread. *Eleutherodactylus johnstonei* (28 islands or island groups), Cuban treefrogs (*Osteopilus septentrionalis*; 14), *Hemidactylus mabouia* (11 and possibly many more), *Gymnophthalmus underwoodi* (15), and *Ramphotyphlops braminus* (12) are small human commensals easily transported inadvertently with goods and ornamental plants. The latter two are relatively inconspicuous and benefit further by being parthenogenetic, thus requiring but a single individual to establish a population. They probably occur on many more islands than have been documented. *Iguana iguana* (15) and *Trachemys scripta* (14) break with the pattern in being large and herbivorous or omnivorous, but both are frequently transported from place to place for food (both historically and recently) or as pets. The situation for *I. iguana*, however, is complicated by the presence of endemic populations that might be subjected to hybridization with more recent arrivals, primarily originating from native Neotropical populations or the introduced populations in Florida and within the region. Two other widely introduced species, *Anolis sagrei* (6) and *Pantherophis guttatus* (11, although established populations have been documented in only a few instances) are notable because of the potential for severe negative consequences resulting from potential competition (*A. sagrei*) with or predation (*P. guttatus*) on native species.

In addition, successful invasions tend to be related to propagule pressure (Kolar and Lodge, 2001; Marchetti et al., 2004). Thus, species that are associated with human economic activity, such as the pet or nursery trade, as well as those with access to frequent commerce-related transport, are more likely to be introduced elsewhere and become established. In our data, the rate of overall arrival was strongly correlated with economic activity, supporting this contention.

Although introductions related to biocontrol are largely relegated to history, movements of animals for food markets (especially *Lithobates catesbeianus* and turtles in the genera *Trachemys* and *Pseudemys*) continue on a large scale. However, in terms of numbers of individuals and species, most alarming is the ever-growing pet trade. The greater Caribbean (especially Haiti) serves as a source of animals, but commercial dealers, particularly in Barbados and the Netherlands Antilles (presumably St. Maarten), undoubtedly pose a greater threat for new introductions into the region. Many of the species moving through the region on their way to and from the US and other markets could easily become established if given the opportunity through escapes or releases. As in Florida (e.g., Meshaka et al., 2004), escapes facilitated by hurricanes and releases, especially by dealers seeking to establish local and easily exploitable populations of popular species, could dramatically change the very nature of the herpetofaunas on a number of islands.

As would be expected from basic principles of island biogeography, the sites subjected to the most introductions tend to be large (or composed of many individual islands), near the sources of many invasives, and/or centers of economic activity. We have documented the most introductions for the Bahamas (32), an archipelago composed of many islands, very close to Florida, and with an active tourist industry. Close behind is Puerto Rico (25), a regional center of commerce, as are Martinique (17), Guadeloupe (16), St.-Martin/St. Maarten (15), and Curaçao (15). The latter also is proximate to the SA mainland, the principal source of introductions to that island. Cuba (9) and Hispaniola (11) are large and have or have had active economic ties to the US.

The success rate (70.3%) for establishing new populations was considerably greater than those calculated by Bomford et al. (2009) for Britain (12 of 51; 23.5%), California (13 of 62; 21.0%), and Florida (47 of 80; 58.8%). Although some of the disparity might be explained by less intensive monitoring in the Caribbean than in Britain or the US, possibly resulting in many colonization attempts going undocumented, much of the success probably is attributable to the hospitable island climates and high incidence of climate-matching with areas where source populations are native (Bomford et al., 2009). The relative paucity of predators, competitors, and pathogens on at least some islands also might contribute to the high rate of successful colonization. In addition, with an increasing number of invasions involving alien species from Florida, the possibility that those species were pre-screened for success by having already established populations outside their native ranges cannot be discounted.

In general, amphibians are less likely than reptiles to successfully colonize islands because of osmotic sensitivity during the dispersal stage and their more stringent ecological requirements during the establishment phase (Vences et al., 2003). In our sample, however, the number of amphibian populations established via human-aided dispersal is sizeable, although the number of reptilian species is considerably greater. The relative abundance of amphibian introductions is largely attributable to four species that are resilient or hardy and have been intentionally introduced or are closely tied to human economic activity. Consequently, species such as *Eleutherodactylus johnstonei*, among the top five most successful colonizing species of amphibians and reptiles (Bomford et al., 2009), *Rhinella marina*, *Scinax* cf. *x-signatus*, and *Osteopilus septentrionalis* are increasingly ubiquitous in the Caribbean and elsewhere.

As additional species become established in the region, and especially in Florida, which remains the source for much of the ornamental vegetation and construction material used in the Caribbean, we will doubtlessly see additional species reported in years to come. These will likely include several other species of *Eleutherodactylus* and *Hemidactylus frenatus*, one of the most widely distributed species in the world (Bomford et al., 2009). Although only a stray *H. frenatus* has been collected on Hispaniola (Scantlebury et al., 2010), a sizeable population appears to be established at the U.S. Naval Base at Guantanamo Bay, Cuba (S. Campbell-Staton, pers. comm.), and it seems only a matter of time before it becomes widely established in the region. The species is highly aggressive and has been successful at displacing similar species (e.g., Powell et al., 1998; Powell, 2004a; Dame and Petren, 2006), raising serious concerns about possible consequences once it arrives in the region. The list of other potential arrivals is long (Kraus, 2009), and several could become serious ecological or economic pests. In addition, some species already in the region, most notably the increasingly widespread green iguana (*I. iguana*; e.g., Sementelli et al., 2008) and the eastern corn snake (*Pantherophis guttatus*), have the potential to become considerably more damaging than they have hitherto been (e.g., van Buurt, 2006, 2011; Platenberg, 2007). Because of the predominance of Florida as a source for invasive amphibians and reptiles, a concerted effort to sanitize cargo and ornamental plants shipped from there is an urgent need.

Although extended dry periods can preclude many unwanted invasives from becoming established, "garden refugia" are available for some species. Amphibians often cannot survive outside of artificially mesic situations (e.g., gardens, golf courses, hotel and resort grounds) during droughts (e.g., *Eleutherodactylus johnstonei* on Anguilla; Hodge et al., 2011). Even some introduced reptilian populations, such as iguanas, are much more plentiful in inhabited areas than in the bush. For snakes, however, this is rarely an option (diminutive and secretive *Ramphotyphlops braminus* might be an exception). Snakes that cannot survive in relatively natural situations and retreat to "gardens" during dry periods find themselves in a "killing zone," where people and domestic predators (dogs and cats) will see them and kill them (Powell and Henderson, 2008). This is very likely why *Pantherophis guttatus* appears to have been extirpated in Curaçao (van Buurt, 2006, 2011).

Although a few attempts have been made to control or eradicate non-native herpetofauna in the Caribbean, such efforts have been rare — and some existing plans have never been implemented. For example, in April 2006, the Ministry of Ecology, Energy and Sustainable Development of Guadeloupe decided to eradicate *Iguana iguana* to prevent competition and hybridization with *Iguana delicatissima*, but nothing was done. Thus, we expect that both firmly and newly established species will generally persist in the region unless policy and management efforts change, causing the greater Caribbean to become part of international trends toward an enhanced pantropical herpetofauna and impoverished native herpetofaunas.

The need to advance protection quickly, perhaps well ahead of political support, flows from the very poor evidence that any environmental Kuznets curve affects these outcomes. An environmental Kuznets curve loosely predicts that, as incomes rise and standards of living improve, greater social support often evolves to mitigate social, environmental, and ecological hazards (Arrow et al., 1995). If this phenomenon holds for invasions in the Caribbean, fig. 1 should begin to display a declining rate of new introductions as GDP rises. That does not seem to be happening at this time. However, economic theory would not predict that invasives would be among the first items corrected as an economy grows. Although growing GDP may have been responsible for declines in emissions of nitrogen oxides, carbon monoxides, sulfur dioxides, and lead in the 1970s and 1980s, the relationship does not seem to hold for aggressive land use conversions to monocultures or impervious surfaces, energy demand, and overall resource consumption. These "high footprint" activities appear to parallel economic development, which would explain why overall atmospheric carbon emissions do not seem to abate with rising GDP (Wagner, 2008). Issues of biodiversity protection in particular have not shown convincing empirical evidence that any abatement turn is emerging on the development horizon (Mills and Waite, 2009). Invasions in the Caribbean would arguably be far behind the curve — or the bend in the curve, as invasions seem to correlate with the very activities most directly responsible for economic growth and development on many islands. Without a much more diverse set of economic activities contributing to economic development, the draw of the US economy and the developments in agriculture, tourism, shipping, and resource extractive industries would seem to continue to accelerate these threats at least for the near and intermediate terms. Precautionary approaches in the name of acute economic stress or intrinsic ecological deterioration from regional invasions are arguably the strongest motivation for the policies suggested.

Although many governmental agencies in the region have addressed invasive species on a case-by-case basis, only the Bahamas has developed and implemented a national invasive species strategy (BEST, 2003). Even there, however, no amphibians or reptiles are listed among the species targeted for eradication or control. The benefits of eradicating an invasive species — a single injection of funds and effort and the problem is solved — far outweigh the cost of a perennial control program (Gardener et al., 2010 and references therein). Many regional introductions remain localized, often in anthropogenic situations (e.g., gardens and grounds of hotels and resorts), and are therefore easy targets for cost-effective eradication projects. Consequently, the development of eradication programs should be a high priority for agencies responsible for managing biodiversity throughout the greater Caribbean.

Nonetheless, prevention remains by far the best — and most economical — approach (Wittenberg and Cock, 2001; Rödder and Weinsheimer, 2010). In that context, increased scrutiny of the transport to and from the islands (whether cargo where inadvertent stowaways may hide, ornamental plants that often carry hitchhikers, or the pet trade that is the source of so many introductions) seems especially desirable. This can help reduce the spread of other problem species, such as agricultural pests, that also are a source of concern for local governments.

To address these concerns, we urge an increased regional and global cooperation on fighting invasive species in general and invasive herpetofauna in particular. Although the region is highly fragmented both geographically and politically, precedents for such cooperation exist; these include the Caribbean Community and Common Market (CARICOM) and the Caribbean Cooperation in Health initiative (www.caricom.org/index.jsp). We urge the adoption of a similarly integrated approach that incorporates not only governmental controls but also investments in local response capacity, such as that advocated by Perry and Farmer (2011). Our combined decades of work in the region show a strong need for considerably more monitoring, education, and research in this area.

Acknowledgements. Support for fieldwork leading to this project was provided by a series of National Science Foundation grants (Research Experiences for Undergraduates) to RP and The Conservation Agency through a grant from the Falconwood Foundation (GP, RP, RWH), by Texas Tech University (GP), Avila University (RP), and the Milwaukee Public Museum and the Windway Foundation (RWH). We are indebted to many governmental agencies that have provided permits to conduct research throughout the greater Caribbean and to the owners of several private islands (Guana, Necker, Little Thatch, and Little St. James islands), many private residences, and hotels or resorts who have tolerated our antics over the years. Facundo Franken, Department of Agriculture and Fisheries in Aruba (DLVVM) and Mikael Lundberg provided information (GVB). Michael E. Seidel helped to clarify our understanding of pond turtles (*Trachemys* spp.) in the Bahamas. This is manuscript T-91202 of the College of Agricultural Sciences and Natural Resource Management, Texas Tech University.

References

Ackley, J.W., Muelleman, P.J., Carter, R.E., Henderson, R.W., Powell, R. (2009): A rapid assessment of herpetofaunal diversity in variously altered habitats on Dominica. Appl. Herpetol. **6**: 171-184.

Alderton, D. (1988): Turtles and Tortoises of the World. New York, Facts on File Publ.

Arias Cornielle, Y.A. (1975): *Anolis chlorocyanus* y *Anolis porcatus* (Sauria, Iguanidae) en la ciudad de Santo Domingo. Unpubl. Tesis, Univ. Autón., Sto. Domingo.

Arrow, K., Bolin, B., Costanza, R., Dasgupta, P., Folke, C., Holling, C. S., Jansson, B.-O., Levin, S., Mäler, K.-G., Perrings, C., Pimentel, D. (1995): Economic growth, carrying capacity, and the environment. Ecol. Econ. **15**: 91-95.

Avila-Pires, T.C.S. (1995): Lizards of Brazilian Amazonia (Reptilia: Squamata). Zool. Verh. Leiden **299**: 1-706.

Ayala, S.C. (1986): Saurios de Colombia: lista actualizada, y distribución de ejemplares colombianos en los museos. Caldasia **15**: 555-575.

Barbour, T. (1910): Notes on the herpetology of Jamaica. Bull. Mus. Comp. Zoöl. **52**: 273-301.

Barbour, T. (1914): A contribution to the zoogeography of the West Indies, with especial reference to amphibians and reptiles. Bull. Mus. Comp. Zoöl. **44**: 209-359.

Barbour, T. (1937): Third list of Antillean reptiles and amphibians. Bull. Mus. Comp. Zool. **82**: 77-166.

Barré, N., Lorvelec, O., Breuil, M. (1997): Les Oiseaux et les Reptiles des Îles de la Petite Terre (Guadeloupe). Bilan d'un Suivi Ecologique d'une Année (Mars 1995 à Mars 1996). Petit-Bourg, Guadeloupe, AEVA, Report No. 16, Office National des Forêts, Conservatoire du Littoral.

Baruš, V., Coy Otero, A. (1974): Nematodes of the genera *Spauligodon*, *Skrjabinodon* and *Pharyngodon* (Oxyuridae) parasitizing Cuban lizards. Věst. Česk. Spol. Zool. **38**: 1-12.

Bayley, I. (1950): The whistling frogs of Barbados. J. Barbados Mus. Hist. Soc. **17**: 161-170.

Berg, C.S., Jeremiah, A., Harrison, B., Henderson, R.W. (2009): New island records for *Tantilla melanocephala* (Squamata: Colubridae) on the Grenada Bank. Appl. Herpetol. **6**: 403-404.

BEST (Bahamas Environment, Science and Technology) Commission. (2003): The National Invasive Species Strategy for the Bahamas. Nassau, BEST.

Bomford, M., Kraus, F., Barry, S.C., Lawrence, E. (2009): Predicting establishment success for alien reptiles and amphibians: A role for climate matching. Biol. Invas. **11**: 713-724.

Boulenger, G.A. (1891): On reptiles, batrachians, and fishes from the Lesser West Indies. Proc. Zool. Soc. London **1891**: 351-357.

Breuil, M. (2000): *Iguana delicatissima* and *Iguana iguana* in FWI. Summer 1999 field report. W. Ind. Iguana Spec. Grp. Newsl. **3**(1): 4-5.

Breuil, M. (2002): Histoire naturelle des amphibiens et reptiles terrestres de l'archipel Guadeloupéen. Guadeloupe, Saint-Martin, Saint Barthélemy. Patr. Nat., Paris **54**: [2] + 339 p.

Breuil, M. (2003): In the footsteps of French naturalists, a "battle" of iguanas, and "improvements" in biodiversity. In: Islands and the Sea. Essays on Herpetological Exploration in the West Indies, p. 255-270. Henderson, R.W., Powell, R., Eds, Ithaca, New York, Soc. Study Amphib. Rept.

Breuil, M. (2004): À la Découverte des Amphibiens & Reptiles des Antilles. Le Gosier, Guadeloupe, PLB (Thierry Petit Le Brun) Édit.

Breuil, M. (2009, 2011): The terrestrial herpetofauna of Martinique: Past, present, future. Appl. Herpetol. **6**: 123-149. Reprinted, with addendum, in: Conservation of Caribbean Island Herpetofaunas, vol. 2, p. 311-338. Hailey, A., Wilson, B.S., Horrocks, J.A., Eds, Leiden, Brill.

Breuil, M., Guiougou, F., Ibéné, B. (2007): Lesser Antillean Iguana (*Iguana delicatissima*). Iguana Spec. Grp. Newsl. **10**(2): 15-17.

Breuil, M., Guiougou, F., Questel, K., Ibéné, B. (2009): Modifications du peuplement herpétologique dans les Antilles françaises: Disparitions et espèces allochtones. 1ère partie: Historique — Amphibiens. Cour. Nat. **249**: 30-37.

Breuil, M., Guiougou, F., Questel, K., Ibéné, B. (2010): Modifications du peuplement herpétologique dans les Antilles françaises: Disparitions et espèces allochtones. 2ème partie: Historique — Reptiles. Cour. Nat. **251**: 36-43.

Breuil, M., Ibéné, B. (2008): Les Hylidés envahissants dans les Antilles francaise et le peuplement batrachologique naturel. Bull. Soc. Herpetol. Fr. **125**: 41-67.

Breuil, M., Sastre, C. (1994): Végétation et reptiles de l'Archipel des Saintes (Antilles françaises). Bull. Amis Mus. Natl. d'Hist. Nat. **178**: 17-19.

Brooks, G.R. (1983): *Gymnophthalmus pleei* Bocourt: An addition to the lizard fauna of Dominica, West Indies. Herpetol. Rev. **14**: 31-32.

Brown, P.R., Echternacht, A.C. (1991): Interspecific behavioral interactions of adult male *Anolis sagrei* and gray-throated *Anolis carolinensis* (Sauria: Iguanidae): A preliminary field study. In: The Fourth Anolis Newsletter, p. 21-30. Losos, J.B., Mayer, G.C., Eds, Washington, D.C., Div. Amphib. Rept., Natl. Mus. Nat. Hist., Smithson. Inst.

Bruner, S.C. (1935): La introducción en Cuba del sapo gigante (*Bufo marinus*). Rev. Agricul. **18**: 73-75.

Buckner, S.D., Franz, R. (1994a): Geographic distribution: *Hemidactylus garnotii*. Herpetol. Rev. **25**: 164.

Buckner, S.D., Franz, R. (1994b): Geographic distribution: *Hemidactylus mabouia*. Herpetol. Rev. **25**: 164.

Buckner, S.D., Franz, R. (1994c): Geographic distribution: *Elaphe guttata*. Herpetol. Rev. **25**: 166.

Buckner, S.D., Franz, R. (1994d): Geographic distribution: *Elaphe obsoleta*. Herpetol. Rev. **25**: 166.

Buckner, S.D., Franz, R. (1998a): Geographic distribution: *Thamnophis sauritus sackenii*. Herpetol. Rev. **29**: 55.

Buckner, S.D., Franz, R. (1998b): Geographic distribution: *Thamnophis sirtalis sirtalis*. Herpetol. Rev. **29**: 55.

Buide, M.S. (1967): Lista de anfibios y reptiles de Cuba. Torreia, n. ser. **1**: 1-60.

Bullock, D.J., Evans, P.G.H. (1990): The distribution, density and biomass of terrestrial reptiles in Dominica, West Indies. J. Zool., London **222**: 421-443.

Bundy, D.A.P., Vogel, P., Harris, E.A. (1987): Helminth parasites of Jamaican anoles (Reptilia: Iguanidae): A comparison of the helminth fauna of 6 *Anolis* species. J. Helminthol. **61**: 77-83.

Burrowes, P.A., Joglar, R.L., Green, D.E. (2004): Potential causes for amphibian declines in Puerto Rico. Herpetologica **60**: 141-154.

Burton, F.J., Echternacht, A.C. (2003): Geographic distribution: *Bufo marinus*. Herpetol. Rev. **34**: 257.

Calsbeek, R. (2009): Experimental evidence that competition and habitat use shape the individual fitness surface. J. Evol. Biol. **22**: 97-108.

Calsbeek, R., Smith, T.B. (2003): Ocean currents mediate evolution in island lizards. Nature **426**: 552-555.

Calsbeek, R., Smith, T.B. (2007): Probing the adaptive landscape on experimental islands: Density dependent selection on lizard body-size. Evolution **61**: 1052-1061.

Calsbeek, R., Bonneaud, C., Smith, T.B. (2008): Differential fitness effects of immunocompetence and neighbourhood density in alternative female lizard morphs. J. Anim. Ecol. **77**: 103-109.

Camargo, A., Heyer, W.R., de Sá, R.O. (2009): Phylogeography of the frog *Leptodactylus validus* (Amphibia: Anura): Patterns and timing of colonization events in the Lesser Antilles. Mol. Phylogen. Evol. **53**: 571-579.

Campbell, D.G. (1978): The Ephemeral Islands: A Natural History of the Bahamas. London, MacMillan Publ.

Carey, E. (2002): Alligator removed from Great Harbor Cay, Bahamas. Croc. Spec. Grp. Newsl. **21**: 12-13..

Carpenter, C.C., Gillingham, J.C. (1984): Giant centipede (*Scolopendra alternans*) attacks marine toad (*Bufo marinus*). Carib. J. Sci. **20**: 71-72.

Carpenter, C.C., Gillingham, J.C. (1987): Water hole fidelity in the marine toad, *Bufo marinus*. J. Herpetol. **21**: 158-161.

Carter, R.E., Berg, C.S., Ackley, J.W., Powell, R. (2009): Frogs of Dominica, with notes on habitat use by two species of *Eleutherodactylus*. Herpetol. Bull. **108**: 14-23.

Censky, E.J. (1988): *Geochelone carbonaria* (Reptilia: Testudines) in the West Indies. Fl. Sci. **51**: 108-114.

Censky, E.J. (1989): *Eleutherodactylus johnstonei* (Salientia: Leptodactylidae) from Anguilla, West Indies. Carib. J. Sci. **25**: 229-230.

Censky, E.J., Hodge, K. (1997): Geographic distribution: *Ramphotyphlops braminus*. Herpetol. Rev. **28**: 210.

Censky, E.J., Kaiser, H. (1999): The Lesser Antillean fauna. In: Caribbean Amphibians and Reptiles, p. 181-221. Crother, B.I., Ed., San Diego, Acad. Press.

Censky, E.J., Lindsay, K. (1997): Geographic distribution: *Gymnophthalmus underwoodi*. Herpetol. Rev. **28**: 210.

Censky, E.J., Hodge, K., Dudley, J. (1998): Over-water dispersal of lizards due to hurricanes. Nature **395**: 556.

Christian, K.A. (1986): Aspects of the life history of Cuban iguanas on Isla Magueyes, Puerto Rico. Carib. J. Sci. **22**: 159-164.

Christian, K.A., Lawrence, W.T. (1991): Microclimate conditions in nests of the Cuban iguana (*Cyclura nubila*). Biotropica **23**: 287-293.

Christian, K.A., Clavijo, I.E., Cordero-López, N., Elias-Maldonado, E.E., Franco, M.A., Lugo-Ramírez, M.V., Marengo, M. (1986): Thermoregulation and energetics of a population of Cuban iguanas (*Cyclura nubila*) on Isla Magueyes, Puerto Rico. Copeia **1986**: 65-69.

Clark, A.H. (1916): The present status and breeding season of the giant toad (*Bufo agua*) in Barbados, St. Vincent, Trinidad and Demerara. Copeia **1916**: 13-14.

Clavero, M., García-Berthou, E. (2005): Invasive species are a leading cause of animal extinctions. Trends Ecol. Evol. **20**: 110.

Cochran, D.M. (1941): The herpetology of Hispaniola. Bull. U.S. Natl. Mus. **177**: 1-398.

Cofresi-Sala, F., Rodríguez de Vega, E. (1963): A new host record for *Mesocoelium danforthi* Hoffman, 1935 (Trematoda: Brachycoeliidae). Carib. J. Sci. **3**: 213.

Cole, C.J. (1975): Evolution of parthenogenetic species of reptiles. In: Intersexuality in the Animal Kingdom, p. 340-355. Reinboth, R., Ed., Berlin, Springer-Verlag.

Cole, C.J., Dessauer, H.C., Townsend, C.R., Arnold, M.G. (1990): Unisexual lizards of the genus *Gymnophthalmus* (Reptilia: Teiidae) in the Neotropics: Genetics, origin, and systematics. Amer. Mus. Novit. **2994**: 1-29.

Conant, R., Collins, J.T. (1991): A Field Guide to the Reptiles and Amphibians of Eastern and Central North America, 3rd edn. Boston, Houghton Mifflin.

Cope, E.D. (1885): Twelfth contribution to the herpetology of tropical America. Proc. Amer. Phil. Soc. **22**: 167-194.

Corke, D. (1992): The status and conservation needs of the terrestrial herpetofauna of the Windward Islands (West Indies). Biol. Conserv. **62**: 47-58.

Corrie, D.G. (2001): A Guide to Barbadian Herpetofauna (Including Introduced Species). St. Michael, First Nations Publ.

Coy Otero, A., Baruš, V. (1979): Nematodes parasitizing Cuban Reptiles. Acta Sci. Nat. Brno **13**: 1-43.

Coy Otero, A., Martínez, J.J. (1987): Nuevo hallazgo en Cuba de larvas del género *Eustrongylides* Jaegerskiöld, 1909 (Nematoda: Dioctophymidae) en la rana toro (*Rana catesbeiana*). Misc. Zool. **32**: 3.

Coy Otero, A., Ventosa, L. (1984): Nemátodos parásitos de anfibios cubanos. Poeyana **269**: 1-20.

Crombie, R.I. (1972): The presence of *Hyla squirella* in the Bahamas. Quart. J. Florida Acad. Sci. **35**: 49-52.

Crombie, R.I. (1999): Jamaica. In: Caribbean Amphibians and Reptiles, p. 63-92. Crother, B.I., Ed., San Diego, Acad. Press.

Crombie, R.I., Steadman, D.W., Barber, J.C. (1984): A preliminary survey of the vertebrates of Cabarita Island, St. Mary Parish, Jamaica. Atoll Res. Bull. **280**: 1-12.

Crother, B.I. (1985): Geographic distribution: *Gastrophryne carolinensis*. Herpetol. Rev. **16**: 114.

Dacosta-Cottam, M., Godbeer, K., Austin, T. (2010): Knight anoles (*Anolis equestris*) on Grand Cayman Island. Rept. & Amphib. **17**: 122.

Daltry, J.C. (2007, 2011): An introduction to the herpetofauna of Antigua, Barbuda and Redonda, with some conservation recommendations. Appl. Herpetol. **4**: 97-130. Reprinted, with addendum, in: Conservation of Caribbean Island Herpetofaunas, vol. 2, p. 17-51. Hailey, A., Wilson, B.S., Horrocks, J.A., Eds, Leiden, Brill.

Dame, E.A., Petren, K. (2006): Behavioural mechanisms of invasion and displacement in Pacific island geckos (*Hemidactylus*). Anim. Behav. **71**: 1165-1173.

Danforth, S.T. (1925): Porto Rican herpetological notes. Copeia **1925**: 76-79.

Daniells, E.A., Ackley, J.W., Carter, R.E., Muelleman, P.J., Rudman, S.M., Turk, P.A., Vélez Espinet, N.J., White, L.A., Wyszynski, N.N. (2008): An annotated checklist of the amphibians and reptiles of Dominica, West Indies. Iguana **15**: 130-141.

Daniells, E.A., Vélez Espinet, N.J., Thorpe, R.S., Powell, R. (2010): *Sphaerodactylus fantasticus*. Cat. Amer. Amphib. Rept. **875**: 1-8.

Daudin, J., de Silva, M. (2007, 2011): An annotated checklist of the amphibians and terrestrial reptiles of the Grenadines with notes on their local natural history and conservation. Appl. Herpetol. **4**: 163-176. Reprinted in: Conservation of Caribbean Island Herpetofaunas, vol. 2, p. 259-271. Hailey, A., Wilson, B.S., Horrocks, J.A., Eds, Leiden, Brill.

Day, M.L., Thorpe, R.S. (1996): Population differentiation of *Iguana delicatissima* and *I. iguana* in the Lesser Antilles. In: Contributions to West Indian Herpetology: A Tribute to Albert Schwartz, p. 436-437. Powell, R., Henderson, R.W., Eds, Contrib. Herpetol., vol. 12. Ithaca, New York, Soc. Study Amphib. Rept.

Day, M.L., Breuil, M., Reichling, S. (2000): Lesser Antillean iguana *Iguana delicatissima*. In: West Indian Iguanas: Status Survey and Conservation Action Plan, p. 62-67. Alberts, A., Ed., Gland, Switzerland, IUCN.

de Armas, L.F., Chirino, N., Armiñana, R., Travieso, J.E. (1987): Macrofauna acuática de la Cueva de Agua, Sagua la Grande, Provincia Villa Clara. Misc. Zool. **34**: 2-3.

Devas, R.P. (1964): History of the Island of Grenada, 1498-1796: With Some Notes and Comments on Carriacou and Events of Later Years. St. George's, Grenada, Carenage Press.

Dexter, R.R. (1932): The food habits of the imported toad *Bufo marinus* in the sugar cane sections of Porto Rico. Proc. 4th Congr. Intl. Soc. Sugar Cane Technol., San Juan **74**: 2-6.

Dickinson, H.C., Fa, J.E. (2000): Abundance, demographics and body condition of a translocated population of St. Lucia Whiptail Lizards (*Cnemidophorus vanzoi*). J. Zool., Lond. **251**: 187-197.

Dodd, C.K., Jr. (1978): Island lizard in danger. Natl. Pks. Conserv. Mag. **52**(8): 10-11.

Duellman, W.E., Crombie, R.I. (1970): *Hyla septentrionalis*. Cat. Amer. Amphib. Rept. **92**: 1-4.

Dukes, J.S., Mooney, H.A. (1999): Does global change increase the success of biological invaders? Trends Ecol. Evol. **14**: 135-139.

Dundee, H.A. (1990): *Anolis* (*Ctenonotus*) *distichus*. Herpetol. Rev. **21**: 22.

Dunn, E.R. (1926): The frogs of Jamaica. Proc. Boston Soc. Nat. Hist. **38**: 111-130.

Dunson, W.A., Seidel, M.E. (1986): Salinity tolerance of estuarine and insular emydid turtles (*Pseudemys nelsoni* and *Trachemys decussata*). J. Herpetol. **20**: 237-245.

Dyer, W.G., Bunkley-Williams, L., Williams, E.H., Jr. (1999): Two new Caribbean records of parasitic nematodes collected from reptiles in Puerto Rico: *Aplectana pusilla* in *Amphisbaena bakeri* and *Alaeuris vogelsangi* in *Iguana iguana*. Carib. J. Sci. **35**: 158-159.

Eaton, J.M., Howard, K.G., Powell, R. (2001): Geographic distribution: *Anolis carolinensis*. Herpetol. Rev. **32**: 118.

Echternacht, A.C., Burton, F.J. (2002): Geographic distribution: *Hemidactylus mabouia*. Herpetol. Rev. **33**: 148.

Echternacht, A.C., Burton, F.J. (2003): Geographic distribution: *Ramphotyphlops braminus*. Herpetol. Rev. **34**: 265-266.

Engeman, R.M., Smith, H.T., Constantine, B. (2005): Invasive green iguanas as airstrike hazards at San Juan International Airport, Puerto Rico. J. Aviat.-Aerosp. Educ. Res. **14**: 45-50.

Ernst, C.H., Barbour, R.W. (1989): Turtles of the World. Washington, DC, Smithson. Inst. Press.

Escobar Herrera, T.R. (1995): Isla de la Juventud: Vertebrados Introducidos por Causas Deliberadas. Habana, Edit. Cien. Técn.

Esteal, S. (1981): The history of introduction of *Bufo marinus* (Amphibia: Anura); a natural experiment in evolution. Biol. J. Linn. Soc. **16**: 93-113.

Esteal, S., Floyd, R.B., Sabath, M.D. (1981): Distribution records of the Marine Toad (*Bufo marinus*). Part 3, The Caribbean. Working Paper no. 5/81. Queensland, School Austral. Environ. Stud., Griffith Univ.

Estrada, A.R., Ruibal, R. (1999): A review of Cuban herpetology. In: Caribbean Amphibians and Reptiles, p. 31-62. Crother, B.I., Ed., California, Acad. Press.

Evans, P. (1989): Herpetofauna of the Commonwealth of Dominica (Windward I., Lesser Antilles), West Indies. Brit. Herpetol. Soc. Bull. **28**: 5-7.

Everard, C.O.R., Sulzer, C.R., Bhagwandin, L.J., Fraser-Champong, G.M., James, A.C. (1980): Pathogenic *Leptospira* isolates from the Caribbean islands of Trinidad, Grenada and St. Vincent. Intl. J. Zoonoses **7**: 90-100.

Everard, C.O.R., Fraser-Champong, G.M., Bhagwandin, L.J., Race, M.W., James, A.C. (1983): Leptospires in wildlife from Trinidad and Grenada. J. Wildl. Diseases **19**: 192-199.

Everard, C.O.R., Carrington, D., Korver, H., Everard, J.D. (1988): Leptospires in the marine toad (*Bufo marinus*) on Barbados. J. Wildl. Diseases **24**: 334-338.

Everard, C.O.R., Carrington, D.G., Korver, H., Burke, R., Everard, J.D., Gravekamp, C. (1990): Leptospires in the whistling frog (*Eleutherodactylus johnstonei*) on Barbados. J. Trop. Med. Hyg. **93**: 140-145.

Feilden, H.W. (1889): The reptiles of Barbados. Zoologist **1889**: 298.

Feilden, H.W. (1903): The reptiles of Barbados. Agricul. News **2**: 365, 381 (reprint of 1889 article).

Fields, A., Horrocks, J.A. (2009): An annotated checklist of the herpetofauna of Barbados. J. Barbados Mus. Hist. Soc. **55**: 263-283.

Fields, A., Horrocks, J.A. (2011): The herpetofauna of Barbados: Anthropogenic impacts and conservation status. In: Conservation of Caribbean Island Herpetofaunas, vol. 2, p. 89-104. Hailey, A., Wilson, B.S., Horrocks, J.A., Eds, Leiden, Brill.

Fitch, H.S., Henderson, R.W., Guarisco, H. (1989): Aspects of the ecology of an introduced anole: *Anolis cristatellus* in the Dominican Republic. Amphib.-Rept. **10**: 307-320.

Fläschendräger, A. (1999): Erstnachweis von *Hemidactylus mabouia* (Moreau de Jonnes, 1818) für Jamaica. Herpetofauna **21**(121): 18.

Fläschendräger, A. (2010): Cuban brown anoles (*Anolis sagrei*) in St. Maarten. Rept. & Amphib. **17**: 121-122.

Forde, A. (2005): Diet of the cane toad *Bufo marinus* in Barbados. The Zool. **13**: 352-353.

Fosberg, F.R. (1983): The human factor in the biogeography of oceanic islands. Comp. Rendu Sci. Soc. Biogéogr. **59**: 147-190.

Franz, R., Morgan, G.S., Davies, J.E. (1987): Some recent introductions of reptiles in the Cayman Islands, West Indies. Herpetol. Rev. **18**: 10-11.

Franz, R., Dodd, C.K., Buden, D.W. (1993): Distributional records of amphibians and reptiles from the Exuma Islands, Bahamas, including the first reports of a freshwater turtle and an introduced gecko. Carib. J. Sci. **29**: 165-173.

Franz, R., Dodd, C.K., Buckner, S.D. (1996): A review of the herpetology of the Bahamian Archipelago. Bahamas J. Sci. **6**: 22-30.

Fritz, U. (1991): Contribution to the knowledge of the Hispaniolan Slider *Trachemys decorata* (Barbour & Carr 1940). Sauria **1**: 11-14.

Garcia, G., Cunningham, A.A., Horton, D.L., Garner, T.W.J., Hyatt, A., Hengstberger, S., Lopez, J., Ogrodowczyk, A., Fenton, C., Fa, J.E. (2007): Mountain chickens *Leptodactylus fallax* and sympatric amphibians appear to be disease free on Montserrat. Oryx **41**: 398-401.

Gardener, M.R., Atkinson, R., Rentería, J.L. (2010): Eradications and people: Lessons from the plant eradication program in Galapagos. Restor. Ecol. **18**: 20-29.

Garrido, O.H., Jaume, M.L. (1984): Catálogo descriptivo de los anfibios y reptiles de Cuba. Doñana, Acta Vert. **11**: 5-128.

Gerber, G.P., Echternacht, A.C. (2000): Evidence for asymmetrical intraguild predation between native and introduced *Anolis* lizards. Oecologia **124**: 599-607.

Germano, J.M., Sander, J.M., Henderson, R.W., Powell, R. (2003): Herpetofaunal communities in Grenada: A comparison of altered sites, with an annotated checklist of Grenadian amphibians and reptiles. Carib. J. Sci. **39**: 68-76.

Giannasi, N., Thorpe, R.S., Malhotra, A. (1997): Introductions of *Anolis* species to the island of St. Lucia, West Indies: Testing for hybrids using multivariate morphometrics. J. Herpetol. **31**: 586-589.

Gifford, M.E., Fontenot, B.E., Nieves, D.M. (2002): Geographic distribution: *Anolis porcatus*. Herpetol. Rev. **33**: 223.

Goin, C.J. (1947): Studies on the life history of *Eleutherodactylus ricordii planirostris* (Cope) in Florida with special reference to the local distribution of an allelomorphic color pattern. Univ. Florida Stud. Biol., Sci. Ser. **4**: 1-67.

Goin, C.J., Cooper, B.W. (1950): Notes on a collection of amphibians from Jamaica. Occ. Pap. Mus. Inst. Jamaica **4**: 1-9.

Goldberg, S.R., Bursey, C.R. (1996): *Anolis maynardi*. Endoparasites. Herpetol. Rev. **27**: 19.

Goldberg, S.R., Bursey, C.R., Tawil, R. (1995): Gastrointestinal helminthes of *Eleutherodactylus johnstonei* (Leptodactylidae) from Bermuda. J. Helminthol. Soc. Washington **62**: 67-69.

Goldberg, S.R., Bursey, C.R., Kaiser, H. (1998a): Gastrointestinal helminthes of five species of *Eleutherodactylus* (Anura: Leptodactylidae) from the West Indies. Carib. J. Sci. **34**: 146-149.

Goldberg, S.R., Bursey, C.R., Cheam, H. (1998b): Helminths of six species of *Anolis* lizards (Polychrotidae) from Hispaniola, West Indies. J. Parasitol. **84**: 1291-1295.

Goodyear, N., Lazell, J. (1994): Status of a relocated population of endangered *Iguana pinguis* on Guana Island, British Virgin Islands. Restor. Ecol. **2**: 43-50.

Gorman, G.C. (1976): Observations on the distribution of *Anolis extremus* (Sauria: Iguanidae) on St. Lucia — A colonizing species. Herpetologica **32**: 184-188.

Gorman, G.C., Kim, Y.J., Yang, S.Y. (1978): The genetics of colonization: Loss of variability among introduced populations of *Anolis* lizards (Reptilia, Lacertilia, Iguanidae). J. Herpetol. **12**: 47-51.

Gosse, P.H. (1850): Descriptions of a new genus and six new species of saurian reptiles. Ann. Mag. Nat. Hist. (2) **6**: 3.

Gosse, P.H. (1851): A Naturalist's Sojourn in Jamaica. London, Longman, Brown, Green, and Longmans.

Grant, C. (1931): Notes on *Bufo marinus* (Linnaeus). Copeia **1931**: 62.

Grant, C. (1932a): The hemidactyls of the Porto Rico region. J. Dept. Agricul. P. Rico **16**: 51-57.

Grant, C. (1932b): Herpetological notes from the Puerto Rico area. J. Dept. Agricul. P. Rico **16**: 161-165.

Grant, C. (1937a): Two early Puerto Rican herpetologists. J. Agricul., Univ. P. Rico **21**: 491-501.

Grant, C. (1937b): Herpetological notes with new species from the American and British Virgin Islands, 1936. J. Agricul. Univ. P. Rico **21**: 503-522.

Grant, C. (1940): The herpetology of the Cayman Islands. Bull. Inst. Jamaica, Sci. Ser. **2**: 1-65.

Grant, C. (1946): Note from Major Chapman Grant, San Diego, California. Nat. Hist. Notes Nat. Hist. Soc. Jamaica **3**: 78.

Grant, C. (1958): A new *Gymnophthalmus* (Reptilia, Teiidae) from Barbados, B.W.I. Herpetologica **14**: 227-228.

Grant, C. (1959): Herpetology of Barbados, B.W.I. Herpetologica **15**: 97-101.

Grant, C., Beatty, H.A. (1944): Herpetological notes on St. Croix, Virgin Islands. Herpetologica **2**: 110-113.

Greene, B.T., Yorks, D.T., Parmerlee, J.S., Jr., Powell, R., Henderson, R.W. (2002): Discovery of *Anolis sagrei* in Grenada with comments on its potential impact on native anoles. Carib. J. Sci. **38**: 270-272.

Greene, B.T., Powell, R., Henderson, R.W. (2003): *Mastigodryas bruesi*. Cat. Amer. Amphib. Rept. **777**: 1-3.

Groombridge, B. (1982): The IUCN Amphibia-Reptilia Red Data Book. Part I. Testudines, Crocodylia, Rhynchocephalia. Gland, Switzerland, IUCN.

Groome, J.R. (1970): A Natural History of the Island of Grenada, West Indies. Trinidad, Carib. Printers.

Haneline, P.G. (1977): Geographic distribution: *Anolis porcatus*. Herpetol. Rev. **8**: 39.

Hardy, J.D., Jr. (1982): Biogeography of Tobago, West Indies, with special reference to amphibians and reptiles: A review. Bull. Maryland Herpetol. Soc. **18**: 37-142.

Hardy, J.D., Jr. (1985): Frog mountain: Preliminary comments on the genus *Eleutherodactylus* on the island of Guadeloupe, West Indies. Bull. Maryland Herpetol. Soc. **21**: 27-33.

Hardy, J.D., Jr., Harris, H.S., Jr. (1979): Occurrence of the West Indian frog, *Eleutherodactylus johnstonei*, in South America and on the island of Curaçao. Bull. Maryland Herpetol. Soc. **15**: 124-133.

Hardy, J.D., Jr., Heyer, W.R., Hedges, S.B., Powell, R. (2004): *Leptodactylus validus*. In: IUCN 2008. 2008 IUCN Red List of Threatened Species (www.iucnredlist.org).

Hardy, L.M., Cole, C.J., Townsend, C.R. (1989): Parthenogenetic reproduction in the Neotropical unisexual lizard, *Gymnophthalmus underwoodi* (Reptilia: Teiidae). J. Morphol. **201**: 215-234.

Hayes, W.K., Carter, R.L., Cyril, S., Thorton, B. (2004): Conservation of an endangered Bahamian rock iguana: I. Population assessments, habitat restoration, and behavioral ecology. In: Iguanas: Biology and Conservation, p. 232-257. Alberts, A., Carter, R., Hayes, W., Martins, E., Eds, Berkeley, Univ. Calif. Presses.

Heatwole, H., De Austin, S.B., Herrero, R. (1968): Heat tolerances of tadpoles of two species of tropical anurans Comp. Biochem. Physiol. **27**: 807-815.

Hedges, S.B. (1996): The origin of West Indian amphibians and reptiles. In: Contributions to West Indian Herpetology: A Tribute to Albert Schwartz, p. 95-128. Powell, R., Henderson, R.W., Eds, Contributions to Herpetology, vol. 12. Ithaca, New York, Soc. Study Amphib. Rept.

Hedges, S.B. (2008): At the lower size limit in snakes: Two new species of threadsnakes (Squamata: Leptotyphlopidae: *Leptotyphlops*) from the Lesser Antilles. Zootaxa **1841**: 1-30.

Hedges, S.B., Incháustegui, S.J., Hernandez, M., Powell, R. (2004): *Osteopilus vastus*. In: IUCN 2010. IUCN Red List of Threatened Species. Version 2010.1 (www.iucnredlist.org).

Henderson, R.W., Berg, C.S. (2005): *Thecadactylus rapicauda*. Diet. Herpetol. Rev. **36**: 454.

Henderson, R.W., Berg, C.S. (2006, 2011): The herpetofauna of Grenada and the Grenada Grenadines: Conservation concerns. Appl. Herpetol. **3**: 197-213. Reprinted, with addendum, in: Conservation of Caribbean Island Herpetofaunas, vol. 2, p. 239-258. Hailey, A., Wilson, B.S., Horrocks, J.A., Eds, Leiden, Brill.

Henderson, R.W., Daudin, J., Haas, G.T., McCarthy, T.J. (1992): Significant distribution records for some amphibians and reptiles in the Lesser Antilles. Carib. J. Sci. **28**: 101-103.

Henderson, R.W., Delatte, A., McCarthy, T.J. (1993): *Gekko gecko* (Sauria: Gekkonidae) established on Martinique, French West Indies. Carib. J. Sci. **29**: 128-129.

Henderson, R.W., Powell, R. (2001). Responses by the West Indian herpetofauna to human-influenced resources. Carib. J. Sci. **37**: 41-54.

Henderson, R.W., Powell, R. (2005): Geographic distribution: *Anolis sagrei*. Herpetol. Rev. **36**: 467.

Henderson, R.W., Powell, R. (2006): Geographic distribution: *Tantilla melanocephala*. Herpetol. Rev. **37**: 501.

Henderson, R.W., Powell, R. (2009): Natural History of West Indian Reptiles and Amphibians. Gainesville, Univ. Press Florida.

Hines, K.N., Iverson, J.B. (2006a): Taxon reports: Allen Cays Iguana (*Cyclura cychlura inornata*). Iguana Spec. Grp. Newsl. **9**(1): 10-11.

Hines, K.N., Iverson, J.B. (2006b): Allen Cays Iguana (*Cyclura cychlura inornata*). Iguana **13**: 279-281.

Hodge, K.V.D., Censky, E.J., Powell, R. (2003): The Reptiles and Amphibians of Anguilla, British West Indies. TheValley, Anguilla, British West Indies, Anguilla Natl. Trust.

Hodge, K.V.D., Powell, R., Censky, E.J. (2011): Conserving the herpetofauna of Anguilla. In: Conservation of Caribbean Island Herpetofaunas, vol. 2, p. 3-15. Hailey, A., Wilson, B.S., Horrocks, J.A., Eds, Leiden, Brill.

Hodsdon, L.A., Pearson, J.F.W. (1943): Notes on the discovery and biology of two Bahaman freshwater turtles of the genus *Pseudemys*. Proc. Florida Acad. Sci. **6**: 17-23.

Hoffman, W.A., Janer, J.L. (1941): El *Bufo marinus* vector de huevos de helmintos en la isla de Puerto Rico. P. Rico J. Pub. Health Trop. Med. **16**: 505-509.

Hoffman, W.H., Noble, G.K. (1927): The bullfrog in Cuba. Copeia **1927**: 59-60.

Horwith, B., Lindsay, K. (1999): A Biodiversity Profile of St. Kitts and Nevis. Revised 2000 by B. Potter, edited by J. Towle. St. John's, Antigua, Island Resource Foundation.

Howard, K.G., Parmerlee, J.S., Jr., Powell, R. (2001): Natural history of the edificarian geckos *Hemidactylus mabouia*, *Thecadactylus rapicauda*, and *Sphaerodactylus sputator* on Anguilla. Carib. J. Sci. **37**: 285-288.

ISSG (Invasive Species Specialist Group). (2010): Global Invasive Species Database. *Python molurus bivittatus* (www.issg.org/database/species/ecology.asp?si=1207&fr=1&sts).

IUCN (International Union for Conservation of Nature and Natural Resources). (2010): 2010 IUCN Red List of Threatened Species (www.iucnredlist.org).

Iverson, J.B. (1992): A Revised Checklist with Distribution Maps of the Turtles of the World. Richmond, Indiana, priv. printed.

Iverson, J.B. (2000): Allen's Cay Iguana, *Cyclura cychlura inornata*. In: West Indian Iguanas: Status Survey and Conservation Action Plan, p. 34-36. Alberts, A., Ed., Gland, Switzerland and Cambridge, UK, IUCN.

Iverson, J.B., Converse, S.J., Smith, G.R., Valiulis, J.M. (2006): Long-term trends in the demography of the Allen Cays Rock Iguana (*Cyclura cychlura inornata*): Human disturbance and density-dependent effects. Biol. Conserv. **132**: 300-310.

Jacobs, J. (1973a): Geographic distribution: *Gastrophryne carolinensis*. HISS News-Journal **1**: 98.

Jacobs, J. (1973b): Geographic distribution: *Rana sphenocephala*. HISS News-Journal **1**: 98.

Jaume, M.L. (1966): Catalogo de la fauna de Cuba. 15. Catalogo de los anfibios de Cuba. Mus. "Felipe Poey" Acad. Cien. Cuba, Trab. Divul. **35**: 1-19.

Jeffrey-Smith, M. (1946): The whistling frog increases. Nat. Hist. Notes Nat. Hist. Soc. Jamaica **3**: 26.

Joglar, R.L. (1998): Los Coquíes de Puerto Rico: Su Historia Natural y Conservación. San Juan, Edit. Univ. P. Rico.

Joglar, R.L., Rios-López, N. (1995): Geographic distribution: *Osteopilus septentrionalis*. Herpetol. Rev. **26**: 105-106.

Joglar, R.L., Rios-López, N. (1998): Geographic distribution: *Eleutherodactylus coqui*. Herpetol. Rev. **29**: 107.

Jones, A.G. (1999): The evolutionary history of *Sphaerodactylus fantasticus*. Unpubl. Ph.D. Thesis, Bangor, UK, Univ. Wales.

Jourdane, J., Theron, A. (1975): Le cycle biologique de *Gorgoderina rochalimai* Pereira et Cuocolo, 1940, digène parasite de *Bufo marinus* en Guadeloupe. Ann. Parasitol. **50**: 439-445.

Kaiser, B.A., Burnett, K.M. (2006): Economic impacts of *E. coqui* frogs in Hawaii. Interdiscip. Environ. Rev. **8**: 1-11.

Kaiser, H. (1992): The trade-mediated introduction of *Eleutherodactylus martinicensis* (Anura: Leptodactylidae) on St. Barthélémy, French Antilles, and its implications for Lesser Antillean biogeography. J. Herpetol. **26**: 264-273.

Kaiser, H. (1997): Origins and introductions of the Caribbean frog, *Eleutherodactylus johnstonei* (Leptodactylidae): Management and conservation concerns. Biodiv. Conserv. **6**: 1391-1407.

Kaiser, H., Hardy, J.D. (1994): *Eleutherodactylus johnstonei*. Cat. Amer. Amphib. Rept. **581**: 1-5.

Kaiser, H., Henderson, R.W. (1994): The conservation status of Lesser Antillean frogs. Herpetol. Nat. Hist. **2**(2): 41-56.

Kaiser, H., Wagenseil, R. (1995): Colonization and distribution of *Eleutherodactylus johnstonei* Barbour (Anura: Leptodactylidae) on Dominica, West Indies. Carib. J. Sci. **31**: 341-344.

Kirby, T. (1986): Return of the monster of the Virgins. BBC Wildl. Mag. **4**(12): 622-623.

Kluge, A.G. (1969): The evolution and geographical origin of the New World *Hemidactylus mabouia-brookii* complex (Gekkonidae, Sauria). Misc. Publ. Mus. Zool. Univ. Michigan **138**: 1-78.

Knapp, C.R. (2002): Exuma Island iguana (*Cyclura cychlura figginsi*), W. Indian Iguana Spec. Grp. Newsl. **5**(1): 7.

Knapp, C.R., Malone, C.L. (2003): Patterns of reproductive success and genetic variability in a translocated iguana population. Herpetologica **59**: 195-202.

Knapp, C.R., Iverson, J.B., Buckner, S.D., Cant, S.V. (2011): Conservation of amphibians and reptiles in the Bahamas. In: Conservation of Caribbean Island Herpetofaunas, vol. 2, p. 53-87. Hailey, A., Wilson, B.S., Horrocks, J.A., Eds, Leiden, Brill.

Knowles, W.C. (1990): Conservation of the St. Croix ground lizard, *Ameiva polops*. Final Report, Endangered species project, study IIB, job IIB 2. St. Croix, Div. Fish Wildl.

Knowles, W.C. (1997): Conservation of the St. Croix ground lizard, *Ameiva polops*. Final Report, Endangered species project, study IIB, job ES 2-1. St. Croix, Div. Fish Wildl.

Kolar, C.S., Lodge, D.M. (2001): Progress in invasion biology: Predicting invaders. Trends Ecol. Evol. **16**: 199-204.

Kolbe, J.J., Glor, R.E., Rodriguez-Schettino, L., Chamizo-Lara, A., Larson, A., Losos, J.B. (2004): Genetic variation increases during biological invasion by a Cuban lizard. Nature **431**: 177-181.

Kolbe, J.J., Glor, R.E., Rodriguez-Schettino, L., Chamizo-Lara, A., Larson, A., Losos, J.B. (2007a): Multiple sources, admixture, and genetic variation in introduced *Anolis* lizard populations. Conservation Biology **21**: 1612-1625.

Kolbe, J.J., Larson, A., Losos, J.B. (2007b): Differential admixture shapes morphological variation among invasive populations of the lizard *Anolis sagrei*. Molecular Ecology **16**: 1579-1591.

Kraus, F. (2009): Alien Reptiles and Amphibians: A Scientific Compendium and Analysis. Invading Nature: Springer Series in Invasion Biology 4. New York, Springer.

Kraus, F., Campbell, E.W. Allison, A., Pratt, T. (1999): *Eleutherodactylus* frog introductions to Hawaii. Herpetol. Rev. **30**: 21-25.

Krysko, K.L., Borgia, A.P. (2005): Geographic distribution: *Hemidactylus mabouia*. Herpetol. Rev. **36**: 468.

Krysko, K.L., King, F.W. (2010): Online Guide to the Snakes of Florida. Gainesville, Florida Mus. Nat. Hist., Univ. Florida (www.flmnh.ufl.edu/herpetology/fl-guide/Boaconstrictor.htm).

Krysko, K.L., Thomas, T.M. (2007): Geographic distribution: *Hemidactylus mabouia*. Herpetol. Rev. **38**: 351.

Landwer, A.J., Ferguson, G.W. (2002): Long-term structural habitat use of male individuals of two native and one introduced *Anolis* (Iguanidae) species on the north coast of Jamaica. Texas J. Sci. **54**: 51-58.

Landwer, A.J., Ferguson, G.W., Herber, R., Brewer, M. (1995): Habitat use of introduced and native anoles (Iguanidae: *Anolis*) along the northern coast of Jamaica. Texas J. Sci. **47**: 45-52.

Lazell, J. (1972): The anoles (Sauria, Iguanidae) of the Lesser Antilles. Bull. Mus. Comp. Zool. **143**: 1-115.

Lazell, J. (1994): A new *Sphaerodactylus* (Sauria: Gekkonidae) from Bequia, Grenada Bank, Lesser Antilles. Breviora **496**: 1-20.

Lazell, J. (1995): Natural Necker. Conserv. Agency Occ. Pap. **2**: 1-28.

Lazell, J. (2000): Necker Island update. IUCN W. Indian Iguana Spec. Grp. Newsl. **3**(1): 2.

Lazell, J. (2002): Restoring vertebrate animals in the British Virgin Islands. Ecol. Restoration **20**: 179-185.

Lazell, J. (2005): Island: Fact and Theory in Nature. Berkeley, Univ. Calif. Press.

Lazell, J. (2006): Natural Necker: Thirty hectares of amazing biological diversity. Conserv. Agency Occ. Pap. **3**: 1-52.

Lazell, J., Sinclair, T. (1990): Geographic distribution: *Eleutherodactylus johnstonei*. Herpetol. Rev. **21**: 95.

Leavitt, B.B. (1933): *Hemidactylus turcicus* in Cuba. Copeia **1933**: 96.

Lee, D.S. (2004): Additional reptiles and amphibians introduced to the Bahamas: A growing conservation concern. Bull. Chicago Herpetol. Soc. **39**: 161-164.

Lee, D.S. (2005): Reptiles and amphibians introduced to the Bahamas; a potential conservation crisis. Bahamas J. Sci. **12**: 2-6.

Lee, D.S., Carey, E. (2001): Conservation concerns facing the Inagua slider. Turt. Tort. Newsl. **3**: 7-9.

Lee, D.S., Ross, J.P. (2001): The Cat Island turtle a reptile of problematic origin; including a bibliography of the genus *Trachemys* in the West Indies region. In: Proc. 8th Symp. Nat. Hist. Bahamas, p. 36-47. Clark-Smith, C., Smith, G., Eds, San Salvador.

Lee, J.C. (1992): *Anolis sagrei* in Florida: Phenetics of a colonizing species III. West Indian and Middle American comparisons. Copeia **1992**: 942-954.

Lemon, R.E. (1971): Vocal communication by the frog *Eleutherodactylus martinicensis*. Can. J. Zool. **49**: 211-217.

Leonard, M.D. (1933): Notes on the giant toad, *Bufo marinus* (L.), in Puerto Rico. J. Econ. Entomol. **26**: 67-72.

Lescure, J. (1966): Le comportement social des batraciens. Rev. Compor. Anim. **2**: 1-33.

Lescure, J. (1979): Singularité et fragilité de la faune en vertebras des Petites Antilles. Comp. Rendu Séances Soc. Biogéogr. **48**: 93-109.

Lescure, J. (1983): Introductions passives et actives de reptiles et d'amphibiens dans les Antilles et les Guyanes. Comp. Rendu Séances Soc. Biogéogr. **59**: 59-70.

Lescure, J. (2000): Répartition passée de *Leptodactylus fallax* Müller, 1923 et d'*Eleutherodactylus johnstonei* Barbour, 1914 (Anoures, Leptodactylidés). Bull. Soc. Herpétol. France **94**: 13-23.

Lescure, J., Marty, C. (1996): Repartition d'*Eleutherodactylus johnstonei* Barbour (Anoure, Leptodactylides), introduction en Guyane Française. Biogeographica (Paris) **72**: 121-125.

Lever, C. (2001): The Cane Toad: The History and Ecology of a Successful Colonist. Otley, Yorkshire, England, Westbury Publ.

Lever, C. (2003): Naturalized Reptiles and Amphibians of the World. New York, Oxford Univ. Press.

Levins, R., Heatwole, H. (1973): Biogeography of the Puerto Rican Bank: Introduction of species onto Palominitos Island. Ecology **54**: 1056-1064.

Long, E.G. (1974): The Serpent's Tale. Reptiles and Amphibians of St. Lucia. Iounaloa Series No. 2. The Morne, St. Lucia, Univ. West Indies Extra Mural Dept.

López-Flores, M., Cruz-Burgos, J.A., Vilella, F.J. (2003): Predation of a white-cheeked pintail (*Anas bahamensis*) duckling by a bullfrog (*Rana catesbeiana*). Carib. J. Sci. **39**: 240-242.

Lorvelec, O., Pascal, M., Pavis, C., Feldmann, P. (2007, 2011): Amphibians and reptiles of the French West Indies: Inventory, threats and conservation. Appl. Herpetol. **4**: 131-161. Reprinted, with addendum, in: Conservation of Caribbean Island Herpetofaunas, vol. 2, p. 205-237. Hailey, A., Wilson, B.S., Horrocks, J.A., Eds, Leiden, Brill.

Losos, J.B. (2009): Lizards in an Evolutionary Tree: Ecology and Adaptive Radiation of *Anolis*. Berkeley and Los Angeles: Univ. Calif. Presses.

Losos, J.B., Marks, J.C., Schoener, T.W. (1993): Habitat use and ecological interactions of an introduced and a native species of *Anolis* lizard on Grand Cayman, with a review of the outcomes of anole introductions. Oecologia **95**: 525-532.

Lundberg, M. (2003): Mellanlandning på Aruba. Snoken **33**: 4-15.

Lynn, W.G. (1937): Two new frogs from Jamaica. Herpetologica **1**: 88-91.

Lynn, W.G. (1940): The herpetology of Jamaica. I. Amphibians. Bull. Inst. Jamaica, Sci. Ser. **1**: 1-60.

Lynn, W.G. (1957): Notes on a collection of reptiles and amphibians from Antigua, B.W.I. Herpetologica **13**: 53-56.

Lynn, W.G., Dent, J.N. (1943): Notes on Jamaican amphibians. Copeia **1943**: 234-242.

Lynn, W.G., Grant, C. (1940): The herpetology of Jamaica. Bull. Inst. Jamaica Sci. Ser. **1**: 1-148.

MacLean, W.P. (1982): Reptiles and Amphibians of the Virgin Islands. London and Basingstoke, Macmillan Caribbean.

Mahon, R., Aiken, K. (1977): The establishment of the North American bullfrog, *Rana catesbeiana* (Amphibia, Anura, Ranidae) in Jamaica. J. Herpetol. **11**: 197-199.

Malhotra, A., Thorpe, R.S. (1999): Reptiles and Amphibians of the Eastern Caribbean. London, MacMillan Educ.

Malhotra, A., Thorpe, R.S., Hypolite, E., James, A. (2007, 2011): A report on the status of the herpetofauna of the Commonwealth of Dominica, West Indies. Appl. Herpetol. **4**: 177-194. Reprinted in: Conservation of Caribbean Island Herpetofaunas, vol. 2, p. 149-166. Hailey, A., Wilson, B.S., Horrocks, J.A., Eds, Leiden, Brill.

Mallery, C.S., Jr., Marcum, M.A., Powell, R., Parmerlee, J.S., Jr., Henderson, R.W. (2007): Herpetofaunal communities of the leeward slopes and coasts of St. Vincent: A comparison of sites variously altered by human activity. Appl. Herpetol. **4**: 313-325.

Malone, C.L., Davis, S.K. (2004): Genetic contributions to Caribbean iguana conservation. In: Iguanas: Biology and Conservation, p. 45-57. Alberts, A., Carter, R., Hayes, W., Martins, E., Eds, Berkeley, Univ. Calif. Presses.

Marchetti, M.P., Moyle, P.B., Levine, R. (2004): Alien fishes in California watersheds: Characteristics of successful and failed invaders. Ecol. Appl. **14**: 587-596.

Marsh, R.E. (1983): Unquestionable evidence that the whistling frog of Barbados is indigenous. J. Barbados Mus. Hist. Soc. **37**: 68-71.

Martínez, J.J., Coy Otero, A., Ventosa, L. (1982): Helmintos de *Rana catesbeiana* Shaw (Ranidae) en Cuba. Poeyana **243**: 1-10.

Martínez, J.L. (1948): Cuban frog leg industry. U.S.F.W.S. Fish. Leafl. **284**: 1-4.

Martínez-Rivera, C.C., González Negron, A., Bertrand, M., Acosta, J. (2003): *Hemidactylus mabouia* (Sauria: Gekkonidae), host of *Geckobia hemidactyli* (Actinedida: Pterygosomatidae), throughout the Caribbean and South America. Carib. J. Sci. **39**: 321-326.

Martins, E.P., Lamont, J. (1998): Estimating ancestral states of a communicative display: A comparative study of *Cyclura* rock iguanas. Anim. Beh. **55**: 1685-1706.

Mausfeld, P., Schmitz, A., Böhme, W., Misof, B., Vrcibradic, D., Rocha, C.F.D. (2002): Phylogenetic affinities of *Mabuya atlantica* Schmidt, 1945, endemic to the Atlantic Ocean Archipelago of Fernando de Noronha (Brazil): Necessity of partitioning the genus *Mabuya* Fitzinger, 1826 (Scincidae. Lygosominae). Zool. Anz. **241**: 281-293.

May, D.W. (1926): The Suriname Toad (*Bufo agua*). Agricul. Notes, P. Rican Exp. Sta. **26**: 2.

May, D.W. (1927): Report of the director. Rep. P. Rico Agricul. Exp. Sta., 1926. Washington, D.C., U.S. Dept. Agricul.

May, D.W. (1930): Report of the director. Rep. P. Rico Agricul. Exp. Sta., 1929. Washington, D.C., U.S. Dept. Agricul.

Mayer, G.C. (1999): Geographic distribution: *Hemidactylus mabouia*. Herpetol. Rev. **30**: 110.

McCoid, M.J. (1995): Non-native reptiles and amphibians. In: Our Living Resources: A Report to the Nation on the Distribution, Abundance, and Health of U.S. Plants, Animals, and Ecosystems, p. 433-437. LaRoe, E.T., Farris, G.S., Puckett, C.E., Doran, P.D., Mac, M.J., Eds, Washington, DC, Natl. Biol. Serv., U.S. Dept. Int.

McCoy, C.J. (1970): *Hemidactylus turcicus*. Cat. Amer. Amphib. Rept. **87**: 1-2.

McGeoch, M.A., Butchart, S.H.M., Spear, D., Marais, E., Kleynhans, E.J., Symes, A., Chanson, J., Hoffmann, M. (2010): Global indicators of biological invasion: Species numbers, biodiversity impact and policy responses. Divers. Distrib. **16**: 95-108.

McKeown, S. (1996): A Field Guide to Reptiles and Amphibians in the Hawaiian Islands. Los Osos, California, Diamond Head Publ.

McManus, J.J., Nellis, D.W. (1975): The critical thermal maximum of the marine toad, *Bufo marinus*. Carib. J. Sci. **15**: 67-70.

McNair, D.B., Mackay, A. (2005): Population estimates and management of *Ameiva polops* (Cope) at Ruth Island, United States Virgin Islands. Carib. J. Sci. **41**: 352-357.

Mealey, B.K., Parks, G.M., Schooley, J., Rothchild, J., Roth, L., Snodgrass, D., Baldwin, J.D. (2002): Discovery of freshwater turtles in Bimini, Bahamas. Bahamas J. Sci. **9**: 58-62.

Mertens, R. (1938): Amphibien und Reptilien aus Santo Domingo, gesammelt von Prof. Dr. H. Böker. Senckenbergiana **20**: 332-342.

Meshaka, W.E., Jr. (1995): Geographic distribution: *Hemidactylus garnotii*. Herpetol. Rev. **26**: 108.

Meshaka, W.E., Jr. (1996): Vagility and the Florida distribution of the Cuban treefrog (*Osteopilus septentrionalis*). Herpetol. Rev. **27**: 37-40.

Meshaka, W.E., Jr. (2001): The Cuban Treefrog in Florida. Life History of a Successful Colonizing Species. Gainesville, Univ. Press Florida.

Meshaka, W.E., Jr., Powell, R. (2009): Diets of the native southern toad (*Anaxyrus terrestris*) and the exotic cane toad (*Rhinella marina*) from a single site in south-central Florida. Florida Sci. **73**: 173-177.

Meshaka, W.E., Jr., Butterfield, B.P., Hauge, J.B. (2004): The Exotic Amphibians and Reptiles of Florida. Malabar, Florida, Krieger.

Metcalf, M.M. (1914): Notes upon *Opalina*. Zool. Anz. **44**: 533-541.

Metcalf, M.M. (1923): The opalinid ciliate infusorians. Bull. U.S. Natl. Mus. **120**: 1-484.

Mettrick, D.F., Dunkley, L.C. (1968): Observations on the occurrence, growth, and morphological variation of the trematode, *Mesocoelium danforthi* Hoffman, 1935, in Jamaica. Carib. J. Sci. **8**: 71-85.

Mills, J.H., Waite, T.A. (2009): Economic prosperity, biodiversity conservation, and the environmental Kuznets curve. Ecol. Econ. **68**: 2087-2095.

Minton, M.R., Minton, S.A. (1975): Geographic distribution: *Hemidactylus mabouia*. Herpetol. Rev. **6**: 116.

Minton, S.A., Minton, M.R. (1984): Geographic distribution: *Anolis sagrei*. Herpetol. Rev. **15**: 77.

Montañez Huguez, L., Sampedro Marín, A., Alvarez Rodríguez, M. (1996): Utilización diferenciada de los recursos espacio y alimento por juveniles y adultos de *Rana catesbeiana* en la Ciénaga de Zapata. Rev. Biol. **10**: 21-26.

Murphy, J.C., Henderson, R.W., Rutherford, M. (2010): *Amphisbaena fuliginosa* (Reptilia: Squamata: Amphisbaenidae) in the Lesser Antilles. Rept. & Amphib. **17**: 181-183.

Nassi, H., Dupouy, J. (1988): Étude expérimentale du cycle biologique d'*Echinostoma parvocirrus* n. sp. (Trematoda: Echinostomatidae), parasite larvaire de *Biomphalaria glabrata* en Guadeloupe. Ann. Parasitol. Hum. Comp. **63**: 103-118.

Nathan, R., Perry, G., Cronin, J.T., Strand, A.E., Cain, M.L. (2003): Methods for estimating long-distance dispersal. Oikos **103**: 261-273.

Neill, W.T. (1964): Frogs introduced on islands. Quart. J. Florida Acad. Sci. **27**: 127-130.

Neils, A., Bugbee, C. (2007): *Rana catesbeiana*. Diet. Herpetol. Rev. **38**: 443.

Norville, A. (2005): The diet of *Bufo marinus* in Barbados. Unpubl. res. proj., Cave Hill, Univ. W. Indies.

Novo-Rodríguez, J., Estrada, A.R., Morena, L.V. (1988): Adiciones a la fauna de anfibios de la Península de Guanahacabibes, Cuba. Misc. Zool. **36**: 3-4.

Odening, K. (1968): Einige Trematoden aus Fröschen and Schildkröten in Vietnam und Kuba. Zool. Anz. **181**: 289-302.

Odum, R.A. (1992): A checklist of the amphibians and reptiles of Aruba. In: International Symposium and Workshop on the Conservation and Research of the Aruba Island Rattlesnake, p. 1-4 (numbered separately). Odum, R.A., Ed., Toledo, Ohio, Toledo Zool. Soc.

Odum, R.A., van Buurt, G. (2009): Geographic distribution: *Anolis porcatus*. Herpetol. Rev. **40**: 450.

Orchard, K. (2010a): *Ramphotyphlops braminus* (Flowerpot Blindsnake). Distribution. Carib. Herpetol. **1**(10): 1.

Orchard, K. (2010b): *Chelonoidis carbonaria* (Red-footed Tortoise). Distribution. Carib. Herpetol. **1**(11): 1.

Orchard, K. (2010c): *Gymnophthalmus underwoodi* (Smooth-scaled Worm Lizard). Distribution. Carib. Herpetol. **1**(12): 1.

Ovaska, K. (1991a): Reproductive phenology, population structure, and habitat use of the frog *Eleutherodactylus johnstonei* in Barbados, West Indies. J. Herpetol. **25**: 424-430.

Ovaska, K. (1991b): Diet of the frog *Eleutherodactylus johnstonei* (Leptodactylidae) in Barbados, West Indies. J. Herpetol. **25**: 486-488.

Ovaska, K. (1992): Short- and long-term movements of the frog *Eleutherodactylus johnstonei* in Barbados, West Indies. Copeia **1992**: 569-573.

Ovaska, K., Hunte, W. (1992): Male mating behavior of the frog *Eleutherodactylus johnstonei* in Barbados, West Indies. Herpetologica **48**: 40-49.

Owen, J.L. (2005): The Cuban tree frog (*Osteopilus septentrionalis*): Distribution, diet, and reproduction of an invasive species in the British Virgin Islands. Unpubl. M.S. Thesis. Lubbock, Texas Tech Univ.

Owen, J.L., Perry, G., Lazell, J., Petrovic, C., Egelhoff, J. (2005a): Geographic distribution: *Osteopilus septentrionalis*. Herpetol. Rev. **36**: 76.

Owen, J.L., Perry, G., Lazell, J., Petrovic, C. (2005b): Geographic distribution: *Pseudemys nelsoni*. Herpetol. Rev. **36**: 466.

Owen, J.L., Perry, G., Lazell, J., Petrovic, C., Egelhoff, J. (2006): *Osteopilus septentrionalis*: Colonization of the British Virgin Islands. Herpetol. Rev. **37**: 74-75.

Paice, M.R. (2005): Geographic distribution: *Bufo marinus*. Herpetol. Rev. **36**: 331-332.

Parmerlee, J.S., Jr., Powell, R., Smith, D.D., Lathrop, A. (1992): Unusual behavior in the Cuban green anole, *Anolis porcatus* (Sauria: Polychridae). Bull. Chicago Herpetol. Soc. **27**: 118.

Pérez, M.E. (1951): The food of *Rana catesbeiana* Show [sic] in Puerto Rico. Herpetologica **7**: 102-104.

Pérez-Buitrago, N.F., Álvarez, A.O., García, M.A. (2006): Cannibalism in an introduced population of *Cyclura nubila nubila* on Isla Magueyes, Puerto Rico. Iguana **13**: 206-208.

Perkins, L.G. (1942): Whistling frogs in St. Ann. Nat. Hist. Notes, Nat. Hist. Soc. Jamaica **4**: 4-5.

Perry, G. (2005): The lizard genus *Anolis*. In: Island: Fact and Theory in Nature, p. 186-190. Lazell, J., Ed., Berkeley, Univ. California Press.

Perry, G. (2009a): First record of *Osteopilus septentrionalis* on Guana Island, British Virgin Islands. Appl. Herpetol. **6**: 191-192.

Perry, G. (2009b): First record of the frog *Eleutherodactylus lentus* in the British Virgin Islands: Conservation implications of native or introduced status. Appl. Herpetol. **6**: 185-187.

Perry, G., Farmer, M. (2011): Reducing the risk of biological invasion by creating incentives for pet sellers and owners to do the right thing. J. Herpetol. **45**: 6-13.

Perry, G., Gerber, G.P. (2006, 2011): Conservation of amphibians and reptiles in the British Virgin Islands: Status and patterns. Appl. Herpetol. **3**: 237-256. Reprinted, with addendum, in: Conservation of Caribbean Island Herpetofaunas, vol. 2, p. 105-127. Hailey, A., Wilson, B.S., Horrocks, J.A., Eds, Leiden, Brill.

Perry, G., Platenberg, R. (2007): Recent additions to the herpetofauna of Little St. James, US Virgin Islands. Appl. Herpetol. **4**: 387-389.

Perry, G., Vice, D. (2009): Forecasting the risk of brown tree snake dispersal from Guam: A mixed transport-establishment model. Conserv. Biol. **23**: 992-1000.

Perry, G., Pierce, J., Griffin, D., van Buurt, G., Lazell, J. (2003): Geographic distribution: *Elaphe guttata guttata*. Herpetol. Rev. **34**: 264.

Perry, G., Powell, R., Watson, H. (2006): Keeping invasive species off Guana Island, British Virgin Islands. Iguana **13**: 272-277.

Perry, G., Owen, J.L., Petrovic, C., Lazell, J., Egelhoff, J. (2007): The red-eared slider, *Trachemy scripta elegans*, in the British Virgin Islands. Appl. Herpetol. **4**: 88-89.

Perry, G., Buchanan, B.W., Fisher, R.N., Salmon, M., Wise, S.E. (2008): Effects of artificial night lighting on reptiles and amphibians in urban environments. In: Urban herpetology, p. 239-256.

Jung, R.E., Mitchell, J.C., Eds, Herpetological Conservation, vol. 3, Salt Lake City, Utah, Soc. Study Amphib. Rept.

Peters, G. (1974): Notizen über der Batrachofauna der Insel Kuba. Mitt. Zool. Mus. Berlin **50**: 299-322.

Philbosian, R., Ruibal, R. (1971): Conservation of the lizard *Ameiva polops* in the Virgin Islands. Herpetologica **27**: 450-454.

Philbosian, R., Yntema, J.A. (1976): Records and status of some reptiles and amphibians in the Virgin Islands. I. 1968-1975. Herpetologica **32**: 81-85.

Philbosian, R., Yntema, J.A. (1977): Annotated Checklist of the Birds, Mammals, Reptiles, and Amphibians of the Virgin Islands and Puerto Rico. Frederiksted, Info. Serv.

Pinchon, R. (1967): Quelques aspects de la nature aux Antilles. Fort-de-France, Martinique: Impr. Caen.

Platenberg, R. (2007): Impacts of introduced species on an island ecosystem: Non-native reptiles and amphibians in the U.S. Virgin Islands. In: Managing Vertebrate Invasive Species: Proceedings of an International Symposium, p. 168-174. Witmer, G.W., Pitt, W.C., Fagerstone, K.A., Eds, Fort Collins, Colorado, USDA/APHIS/WS, Natl. Wildl. Res. Ctr.

Platenberg, R.J., Boulon, R.H., Jr. (2006, 2011): Conservation status of reptiles and amphibians in the U.S. Virgin Islands. Appl. Herpetol. **3**: 215-235. Reprinted, with addendum, in: Conservation of Caribbean Island Herpetofaunas, vol. 2, p. 407-428. Hailey, A., Wilson, B.S., Horrocks, J.A., Eds, Leiden, Brill.

Pough, F.H., Stewart, M.M., Thomas, R.G. (1977): Physiological basis of habitat partitioning in Jamaican *Eleutherodactylus*. Oecologia **27**: 285-293.

Powell, R. (1990a): *Anolis porcatus*. Behavior. Herpetol. Rev. **21**: 93.

Powell, R. (1990b): Geographic distribution: *Anolis porcatus*. Herpetol. Rev. **21**: 96.

Powell, R. (1990c): *Hemidactylus palaichthus*. Cat. Amer. Amphib. Rept. **468**: 1.

Powell, R. (1992): *Anolis porcatus*. Cat. Amer. Amphib. Rept. **541**: 1-5.

Powell, R. (2002): Geographic distribution: *Anolis carolinensis*. Herpetol. Rev. **33**: 321.

Powell, R. (2003): Exploitation of reptiles in the West Indies: A long history. Iguana **10**: 67-70.

Powell, R. (2004a): Species profile: Asian House Gecko (*Hemidactylus frenatus*). Iguana **11**: 20.

Powell, R. (2004b): Conservation of iguanas (*Iguana delicatissima* and *I. iguana*) in the Lesser Antilles. Iguana **11**: 238-246.

Powell, R. (2006, 2011): Conservation of the herpetofauna on the Dutch Windward Islands: St. Eustatius, Saba, and St. Maarten. Appl. Herpetol. **3**: 293-306. Reprinted, with addendum, in: Conservation of Caribbean Island Herpetofaunas, vol. 2, p. 189-204. Hailey, A., Wilson, B.S., Horrocks, J.A., Eds, Leiden, Brill.

Powell, R. (2007): Geographic distribution: *Osteopilus septentrionalis*. Herpetol. Rev. **38**: 215.

Powell, R., Henderson, R.W. (1999): Addenda to the checklist of West Indian amphibians and reptiles. Herpetol. Rev. **30**: 137-139.

Powell, R., Henderson, R.W. (2003): A second set of addenda to the checklist of West Indian amphibians and reptiles. Herpetol. Rev. **34**: 341-345.

Powell, R., Henderson, R.W. (2005): Conservation status of Lesser Antillean reptiles. Iguana **12**: 62-77.

Powell, R., Henderson, R.W. (2007, 2011): The St. Vincent (Lesser Antilles) herpetofauna: Conservation concerns. Appl. Herpetol. **4**: 295-312. Reprinted in: Conservation of Caribbean Island Herpetofaunas, vol. 2, p. 359-376. Hailey, A., Wilson, B.S., Horrocks, J.A., Eds, Leiden, Brill.

Powell, R., Henderson, R.W. (2008): Urban herpetology in the West Indies. In: Urban Herpetology, p. 389-404. Mitchell, J.C., Jung Brown, R.E., Bartholomew, B., Eds, Herpetological Conservation, vol. 3, Salt Lake City, Soc. Study Amphib. Rept.

Powell, R., Incháustegui, S.J. (2009, 2011): Conservation of the herpetofauna of the Dominican Republic. Appl. Herpetol. **6**: 103-122. Reprinted, with addendum, in: Conservation of Caribbean

Island Herpetofaunas, vol. 2, p. 167-188. Hailey, A., Wilson, B.S., Horrocks, J.A., Eds, Leiden, Brill.

Powell, R., Lindsay, K. (1999): Geographic distribution. *Gymnophthalmus underwoodi*. Herpetol. Rev. **30**: 110.

Powell, R., Maxey, S.A. (1990): *Hemidactylus brookii*. Cat. Amer. Amphib. Rept. **493**: 1-3.

Powell, R., Parmerlee, J.S., Jr. (1991): Geographic distribution: *Anolis porcatus*. Herpetol. Rev. **22**: 65.

Powell, R., Smith, D.D., Parmerlee, J.S., Jr., Taylor, C.V., Jolley, M.L. (1990): Range expansion by an introduced anole: *Anolis porcatus* in the Dominican Republic. Amphib.-Rept. **11**: 421-425.

Powell, R., Passaro, R.J., Henderson, R.W. (1992): Noteworthy herpetological records from Saint [*sic*] Maarten, Netherlands Antilles. Carib. J. Sci. **28**: 234-235.

Powell, R., Henderson, R.W., Adler, K., Dundee, H.A. (1996): An annotated checklist of West Indian amphibians and reptiles. In: Contributions to West Indian Herpetology: A Tribute to Albert Schwartz, p. 51-93 + 8 pls. Powell, R., Henderson, R.W., Eds, Contributions to Herpetology, vol. 12. Ithaca, New York, Soc. Study Amphib. Rept.

Powell, R., Crombie, R.I., Boos, H.E.A. (1998): *Hemidactylus mabouia*. Cat. Amer. Amphib. Rept. **674**: 1-11.

Powell, R., Ottenwalder, J.A., Incháustegui, S.J. (1999): The Hispaniolan herpetofauna: Diversity, endemism, and historical perspectives, with comments on Navassa Island. In: Caribbean Amphibians and Reptiles, p. 93-168. Crother, B.I., Ed., San Diego, Acad. Press.

Powell, R., Ottenwalder, J.A., Incháustegui, S.J., Henderson, R.W., Glor, R.E. (2000): Amphibians and reptiles of the Dominican Republic: Species of special concern. Oryx **34**: 118-128.

Powell, R., Henderson, R.W., Parmerlee, J.S., Jr. (2005): Reptiles and Amphibians of the Dutch Caribbean: St. Eustatius, Saba, and St. Maarten. Gallows Bay, St. Eustatius, Netherlands Antilles, St. Eustatius Natl. Pks. Found.

Pregill, G.K., Steadman, D.W., Watters, D.R. (1994): Late Quaternary vertebrate faunas of the Lesser Antilles: Historical components of Caribbean biogeography. Bull. Carnegie Mus. Nat. Hist. **30**: 1-51.

Pritchard, P.C.H. (1979): Encyclopedia of Turtles. Neptune, New Jersey, T.F.H. Publ.

Pritchard, P.C.H., Trebbau, P. (1984): The Turtles of Venezuela. Contrib. Herpetol., vol. 2. Ithaca, New York, Soc. Study Amphib. Rept.

Proctor, G.R. (1973): A saga of frogs. Jamaica J. **7**(4): 29-31.

Pyšek, P., Jarošík, V., Hulme, P.E., Kühn, I., Wild, J., Arianoutsou, M., Bacher, S., Chiron, F., Didžiulis, V., Essl, F., Genovesi, P., Gherardi, F., Hejda, M., Kark, S., Lambdon, P.W., Desprez-Loustau, M.-L., Nentwig, W., Pergl, J., Poboljšaj, K., Rabitsch, W., Roques, A., Roy, D.B., Shirley, S., Solarz, W., Vilà, M., Winter, M. (2010): Disentangling the role of environmental and human pressures on biological invasions across Europe. Proc. Natl. Acad. Sci. (published ahead of print 7 June 2010, doi:10.1073/pnas.1002314107).

Quick, J.S., Reinert, H.K., de Cuba, E.R., Odum, R.A. (2005): Recent occurrence and dietary habits of *Boa constrictor* on Aruba, Dutch West Indies. J. Herpetol. **39**: 304-307.

Reed, R.N., Krysko, K.L., Snow, R.W., Rodda, G.H. (2010): Is the Northern African Python (*Python sebae*) established in southern Florida? Rept. & Amphib. **17**: 52-54.

Reynolds, R.G. (2011): Status, conservation, and introduction of amphibians and reptiles in the Turks and Caicos Islands, British West Indies. In: Conservation of Caribbean Island Herpetofaunas, vol. 2, p. 377-406. Hailey, A., Wilson, B.S., Horrocks, J.A., Eds, Leiden, Brill.

Reynolds, R.G., Niemiller, M.L. (2010): Island invaders: Invasive reptiles and amphibians in the Turks and Caicos Islands. Rept. & Amphib. **17**: 116-121.

Richardson, D.M., Pyšek, P. (2006): Plant invasions: Merging the concepts of species invasiveness and community invisibility. Progr. Phys. Geogr. **30**: 409-431.

Rios-López, N. (1999): Geographic distribution. *Scinax rubra*. Herpetol. Rev. **30**: 232.

Rios-López, N. (2000): Geographic distribution. *Scinax rubra*. Herpetol. Rev. **31**: 51.

Rios-López, N., Joglar, R.L. (1999): Geographic distribution. *Rana grylio*. Herpetol. Rev. **30**: 231-232.

Rivero, J.A. (1978): Los Anfibios y Reptiles de Puerto Rico. The Amphibians and Reptiles of Puerto Rico. San Juan, Edit. Univ. Puerto Rico.

Rivero, J.A. (1998): Los Anfibios y Reptiles de Puerto Rico. The Amphibians and Reptiles of Puerto Rico, 2nd ed. San Juan, Edit. Univ. Puerto Rico.

Rivero, J.A., Joglar, R. (1979): Geographic distribution: *Eleutherodactylus cochranae*. Herpetol. Rev. **10**: 101.

Rödder, D., Weinsheimer, F. (2010): Will future anthropogenic climate change increase the potential distribution of the alien invasive Cuban treefrog (Anura: Hylidae)? J. Nat. Hist. **43**: 1207-1217.

Rodríguez Schettino, L.R. (2000): Cuban reptiles: Original citations, holotypes, and geographic range. Smithson. Herpetol. Info. Serv. **125**: 1-26.

Ross, J.P. (1982): The Cat Island turtle. Oryx **16**: 349-351.

Roughgarden, J., Pacala, S., Rummel, J. (1984): Strong present-day competition between the *Anolis* lizard populations of St. Maarten (Neth. Antilles). Evol. Ecol. **23**: 203-220.

Rueda-Almonacid, J.V. (1998): Aspectos generales sobre la situación actual de las poblaciones adventicias de rana toro (*Rana catesbeiana*) en el Valle del Cauca. In: Informe Nacional sobre el Estado de la Biodiversidad Colombia-1997. Causas de Pérdida de Biodiversidad, vol. II, p. 101-103. Chaves, M.E., Arango, N., Eds, Bogotá, Inst. Invest. Recurs. Biol. Alexander von Humboldt, PNUMA, Min. Medio Ambiente.

Rueda-Almonacid, J.V. (1999): Situación actual y problemática generada por la introducción de "rana toro" a Colombia. Rev. Acad. Colom. Cien. Exac., Fis. Nat. **23** (supl. esp.): 367-393.

Sampedro Marín, A., Montañez Huguez, L. (1993): Especiación incipiente en *Rana catesbeiana* (Ranidae: Anura) de la Cienaga de Zapata, Cuba. Rev. Biol., Univ. Habana **7**: 113-124.

Sampedro Marín, A., Montañez Huguez, L., Suárez Boado, O. (1985): Alimentación de *Rana catesbeiana* en dos zonas de captura de Cuba. Cien. Biol. **13**: 59-66.

Sampedro Marín, A., Montañez Huguez, L., Suárez Boado, O. (1986): Food of *Rana catesbeiana* in two different areas of Cuba. In: Studies in Herpetology. Proceedings of the European Herpetological Meeting, Prague 1985, p. 413-416. Roãek, Z., Ed., Prague, Charles Univ.

Sampedro Marin, A., Rivalta, R.V., Díaz, B.L.M. (2003): Ranas acuáticas y ribereñas. In: Anfibios y Reptiles de Cuba, p. 38-43. Rodríguez Schettino, L., Ed., Vaasa, Finland, UPC Print.

Sander, J.M., Germano, J.M., Powell, R., Henderson, R.W. (2003): Colour and pattern polymorphism in *Eleutherodactylus johnstonei* on Grenada. Herpetol. Bull. **83**: 22-25.

Savage, J.M. (1954): Notulae herpetologicae 1-7. Trans. Kansas Acad. Sci. **57**: 326-334.

Scantlebury, D., Ng, J., Landestoy, M., Glor, R.E. (2010): *Hemidactylus frenatus* and *Gymnophthalmus underwoodi* in the Dominican Republic. Rept. & Amphib. **17**: 180-181.

Schall, J.J. (1973): Relations among three macroteiid lizards on Aruba Island. J. Herpetol. **7**: 289-295.

Schloegel, L.M., Picco, A.M., Kilpatrick, A.M., Davies, A.J., Hyatt, A.D., Daszak, P. (2009): Magnitude of the US trade in amphibians and presence of *Batrachochytrium dendrobatidis* and ranavirus infection in imported North American bullfrogs (*Rana catesbeiana*). Biol. Conserv. **142**: 1420-1426.

Schoener, T.W., Schoener, A. (1971): Structural habitats of West Indian *Anolis* lizards. I. Lowland Jamaica. Breviora **368**: 1-53.

Schoener, T.W., Spiller, D.A. (1996): Devastation of prey diversity by experimentally introduced predators in the field. Nature **381**: 691-694.

Schoener, T.W., Spiller, D.A., Losos, J.B. (2001): Natural restoration of a species-area relation for a lizard after a hurricane. Science **294**: 1525-1528.

Schoener, T.W., Spiller, D.A., Losos, J.B. (2002): Predation of a common *Anolis* lizard: Can the food-web effects of a devastating predator be reversed? Ecol. Monogr. **72**: 383-407.

Schoener, T.W., Losos, J.B., Spiller, D.A. (2005): Island biogeography of populations: An introduced species transforms survival patterns. Science **310**: 1807-1809.

Schomburgk, R.H. (1848): The History of Barbados. London, Frank Cass (reprint 1971).

Schwartz, A. (1965): Geographic variation in *Sphaerodactylus notatus* Baird. Rev. Biol. Trop. **13**: 161-185.

Schwartz, A. (1967): Frogs of the genus *Eleutherodactylus* in the Lesser Antilles. Stud. Fauna Curaçao Other Carib. Isl. **24**: 1-62.

Schwartz, A. (1968): The geckos (*Sphaerodactylus*) of the southern Bahama Islands. Ann. Carnegie Mus. **39**: 227-271.

Schwartz, A. (1970): *Sphaerodactylus notatus*. Cat. Amer. Amphib. Rept. **90**: 1-2.

Schwartz, A. (1972): The native toads (Anura, Bufonidae) of Cuba. J. Herpetol. **6**: 217-231.

Schwartz, A. (1973): *Sphaerodactylus*. Cat. Amer. Amphib. Rept. **142**: 1-2.

Schwartz, A. (1974): *Eleutherodactylus planirostris*. Cat. Amer. Amphib. Rept. **154**: 1-4.

Schwartz, A., Carey, W.M. (1977): Systematics and evolution in the West Indian iguanid genus *Cyclura*. Stud. Fauna Curaçao Other Carib. Isl. **53**: 15-97.

Schwartz, A., Fowler, D.C. (1973): The Anura of Jamaica: A progress report. Stud. Fauna Curaçao Other Carib. Isl. **43**: 50-142.

Schwartz, A., Henderson, R.W. (1985): A Guide to the Identification of the Amphibians and Reptiles of the West Indies Exclusive of Hispaniola. Milwaukee, Milwaukee Pub. Mus.

Schwartz, A., Henderson, R.W. (1988): West Indian amphibians and reptiles: A check-list. Milwaukee Pub. Mus. Contrib. Biol. Geol. **74**: 1-264.

Schwartz, A., Henderson, R.W. (1991): Amphibians and Reptiles of the West Indies: Descriptions, Distributions, and Natural History. Gainesville, University of Florida Press.

Schwartz, A., Thomas, R. (1975): A checklist of West Indian amphibians and reptiles. Carnegie Mus. Nat. Hist. Spec. Publ. **1**: 1-216.

Schwartz, A., Thomas, R., Ober, L.D. (1978): First supplement to a check-list of West Indian amphibians and reptiles. Carnegie Mus. Nat. Hist. Spec. Publ. **5**: 1-35.

Scott, N.J., Jr., Ayala, S.C. (1984): Geographic distribution: *Tretioscincus bifasciatus*. Herpetol. Rev. **15**: 21.

Seidel, M.E. (1988): Revision of the West Indian emydid turtles (Testudines). Amer. Mus. Novit. **2918**: 1-41.

Seidel, M.E. (1990): Growth and population characteristics of the slider turtle, *Trachemys decussata*, on Grand Cayman Island. J. Herpetol. **24**: 191-196.

Seidel, M.E. (1996): Current status of biogeography of the West Indian turtles in the genus *Trachemys* (Emydidae). In: Contributions to West Indian Herpetology: A Tribute to Albert Schwartz, p. 169-174. Powell, R., Henderson, R.W., Eds, Contrib. Herpetol., vol. 12. Ithaca, New York, Soc. Study Amphib. Rept.

Seidel, M.E. (2003): *Trachemys decussata angusta*. Longevity. Herpetol. Rev. **34**: 363-364.

Seidel, M.E., Adkins, M.D. (1987): Biochemical comparisons among West Indian *Trachemys* (Emydidae: Testudines). Copeia **1987**: 485-489.

Seidel, M.E., Ernst, C.H. (2006): *Trachemys scripta*. Cat. Amer. Amphib. Rept. **831**: 1-94.

Seidel, M.E., Franz, R. (1994): Amphibians and reptiles (exclusive of marine turtles) of the Cayman Islands. In: The Cayman Islands: Natural History and Biogeography, p. 407-433. Brunt, M.A., Davis, J.E., Eds, Dordrecht, The Netherlands, Kluwer.

Seín, F., Jr. (1937): The development of the giant Suriname toad, *Bufo marinus* L. J. Agricul. Univ. P. Rico **21**: 77-78.

Sementelli, A., Smith, H.T., Meshaka, W.E., Jr., Engeman, R.M. (2008): Just green iguanas?: The associated costs and policy implications of exotic invasive wildlife in south Florida. Pub. Wks. Mgmt. Pol. **12**: 599-606.

Servan, J., Arvy, C. (1997): Introduction de la tortue de Floride *Trachemys scripta* en France: un nouveau compétiteur pour les espèces de tortues d'eau douce européennes. Bull. Franc. Pêche Piscicult. **344/345**: 173-177.

Snow, R.W., Krysko, K.L., Enge, K.M., Oberhofer, L., Warren-Bradley, A., Wilkins, L. (2007): Introduced populations of *Boa constrictor* (Boidae) and *Python molurus bivittatus* (Pythonidae) in southern Florida. In: Biology of the Boas and Pythons, p. 416-438. Henderson, R.W., Powell, R., Eds, Eagle Mountain, Utah, Eagle Mountain Publ.

Soberon, R., Ramos, R., McMahan, W., Ross, P. (1996): Cuba: Reproduction of Cuban Crocodile on the Isla of Pines. Croc. Spec. Grp. Newsl. **15**(3): 10-11.

Spiller, D.A., Losos, J.B., Schoener, T.W. (1998): Impact of a catastrophic hurricane on island populations. Science **281**: 695-697.

Stewart, M.M. (1977): The role of introduced species in a Jamaican frog community. Act. IV Simp. Intl. Ecol. Trop. **1**: 110-146.

Stewart, M.M., Martin, G.E. (1980): Coconut husk-piles — a unique habitat for Jamaican terrestrial frogs. Biotropica **12**: 107-116.

Strauss, S.Y., Webb, C.O., Salamin, N. (2006): Exotic taxa less related to native species are more invasive. Proc. Natl. Acad. Sci. **103**: 5841-5845.

Thomas, R. (1966): New species of Antillean *Eleutherodactylus*. Quart. J. Florida Acad. Sci. **28**: 375-391.

Thomas, R. (1975): The *argus* group of West Indian *Sphaerodactylus* (Sauria: Gekkonidae). Herpetologica **31**: 177-195.

Thomas, R. (1999): The Puerto Rico area. In: Caribbean Amphibians and Reptiles, p. 169-179. Crother, B.I., Ed., San Diego, Acad. Press.

Thomas, R., Joglar, R. (1996): The herpetology of Puerto Rico: Past, present, and future. Ann. New York Acad. Sci. **776**: 181-200.

Thompson, G.B. (1950): Ticks of Jamaica, B.W.I. — Records and notes (including a summary of the distribution of the West Indian species). Ann. Mag. Nat. Hist., ser. 12, **3**: 220-229.

Thorpe, R.S., Jones, A.G., Malhotra, A., Surget-Groba, Y. (2008): Adaptive radiation in Lesser Antillean lizards: Molecular phylogenetics and species recognition in the Lesser Antillean dwarf gecko complex, *Sphaerodactylus fantasticus*. Mol. Ecol. **17**: 1489-1504.

Tolson, P.J., Henderson, R.W. (2006, 2011): An overview of snake conservation in the West Indies. Appl. Herpetol. **3**: 345-356. Reprinted, with addendum, in: Conservation of Caribbean Island Herpetofaunas, vol. 1, p. 49-62. Hailey, A., Wilson, B.S., Horrocks, J.A., Eds, Leiden, Brill.

Townsend, J.H., Eaton, J.M. Powell, R., Parmerlee, J.S., Jr., Henderson, R.W. (2000): Cuban Treefrogs (*Osteopilus septentrionalis*) in Anguilla, Lesser Antilles. Carib. J. Sci. **36**: 326-328.

Trakhtenbrot, A., Nathan, R., Perry, G., Richardson, D.M. (2005): The importance of long-distance dispersal in conservation. Divers. Distrib. **11**: 173-181.

Treglia, M.L. (2006): An annotated checklist of the amphibians and reptiles of St. Vincent, West Indies. Iguana **13**: 251-262.

Treglia, M.L., Muensch, A.J., Powell, R., Parmerlee, J.S., Jr. (2008): Invasive *Anolis sagrei* on St. Vincent and its potential impact on perch heights of *Anolis trinitatis*. Carib. J. Sci. **44**: 251-256.

Tucker, R.W.E. (1940): *Bufo marinus* L. in Barbados. Agricul. J. **8**: 145-150.

Tucker, R.W.E., Wolcott, G.N. (1935): Parasite introductions: Barbados and Puerto Rico. Proc. 5th Congr. Intl. Soc. Sugar Cane Technol., p. 398-404. Brisbane.

Turk, P.A., Wyszynski, N.N., Powell, R., Henderson, R.W. (2010): Population densities and water-loss rates of *Gymnophthalmus pleii*, *Gymnophthalmus underwoodi* (Gymnophthalmidae), and *Sphaerodactylus fantasticus fuga* (Sphaerodactylidae) on Dominica, West Indies. Salamandra **46**: 125-130.

Underwood, G. (1959): The anoles of the eastern Caribbean (Sauria, Iguanidae). Part III. Revisionary notes. Bull. Mus. Comp. Zool. **121**: 191-226.

Underwood, G. (1962): Reptiles of the eastern Caribbean. Carib. Aff., n. ser. **1**: iv + 192 p.

Underwood, G., Horrocks, J.A., Daltry, J.C. (1999): A new snake from Barbados. J. Barbados Mus. Hist. Soc. **45**: 67-75.

Underwood, G., Williams, E.E. (1959): The anoline lizards of Jamaica. Bull. Inst. Jamaica, Sci. Ser. **9**: 1-48.

USFWS (U.S. Fish & Wildlife Service). (2010): Proposed Rule for Listing Giant Constrictor Snakes as Injurious Wildlife (www.fws.gov/verobeach/index.cfm?method=activityhighlights&id=11).

van Buel, H., Powell, R. (2006): Geographic distribution: *Gymnophthalmus underwoodi*. Herpetol. Rev. **37**: 494.

van Buurt, G. (2001): De Amfibieën en Reptielen van Aruba, Curaçao en Bonaire. Curaçao, priv. publ.

van Buurt, G. (2005): Field Guide to the Amphibians and Reptiless of Aruba, Curaçao and Bonaire. Frankfurt a/Main, Edit. Chimaira.

van Buurt, G. (2006, 2011): Conservation of amphibians and reptiles in Aruba, Curaçao and Bonaire. Appl. Herpetol. **3**: 307-321. Reprinted, with addendum, in: Conservation of Caribbean Island Herpetofaunas, vol. 1, p. 145-159. Hailey, A., Wilson, B.S., Horrocks, J.A., Eds, Leiden, Brill.

van Buurt, G. (2007): Breeding population of *Osteopilus septentrionalis* in Curaçao. Appl. Herpetol. **4**: 390-391.

van Volkenberg, H.L. (1935): Biological control of an insect pest by a toad. Science **82**: 278-279.

van Wijngaarden, R. (1988): Enige opmerkingen over de herpetofauna van Curaçao. Lacerta **46**: 188-193.

Vance, T. (1987): The Cuban green anole (*Anolis porcatus*): A colonizing species. Bull. Maryland Herpetol. Soc. **23**: 105-108.

Vanzolini, P.E. (1978): On South American *Hemidactylus*. Pap. Avul. Zool., São Paulo **31**: 307-343.

Vargas-Salinas, F. (2005): *Bufo marinus*. Amplexus displacement. Herpetol. Rev. **36**: 431-432.

Vargas-Salinas, F. (2006a): *Osteopilus septentrionalis*. Reproduction. Herpetol. Rev. **37**: 205.

Vargas-Salinas, F. (2006b): Sexual size dimorphism in the Cuban treefrog *Osteopilus septentrionalis*. Amphib.-Rept. **27**: 419-426.

Vargas-Salinas, F. (2006c): Breeding behavior and colonization success of the Cuban treefrog *Osteopilus septentrionalis*. Herpetologica **62**: 398-408.

Varona, L.S. (1976): *Caiman crocodilus* (Reptilia: Alligatoridae) en Cuba. Misc. Zool. **5**: 2.

Varona, L.S. (1980): Protection in Cuba. Oryx **15**: 282-284.

Varona, L.S. (1981): Cuban crocodiles. Rephiberary **40**: 7-8.

Vences, M., Vieites, D.R., Glaw, F., Brinkmann, H., Kosuch, J., Veith, M., Meyer, A. (2003): Multiple overseas dispersal in amphibians. Proc. R. Soc. Lond. B **270**: 2435-2442.

Vogel, Z. (1965): Herpetologische Beobachtungen auf Kuba (IV). Froschfang auf Kuba. Aquar. Terr. **12**: 420-422 (reprinted in Dutch as Herpetologische waarnemingen op Cuba V [slot]. Lacerta **25**: 5-7).

Waddle, J.H., Crockett, M.E., Rice, K.G. (2005): Geographic distribution: *Osteopilus septentrionalis*. Herpetol. Rev. **36**: 333.

Waddle, J.H., Crockett, M.E., Rice, K.G. (2006): Geographic distribution: *Eleutherodactylus coqui*. Herpetol. Rev. **37**: 487.

Wagenaar Hummelinck, P. (1940): Studies on the fauna of Curaçao, Aruba, Bonaire and the Venezuelan islands: No. 2. A survey of the mammals, lizards and mollusks. Stud. Fauna Curaçao Other Carib. Isl. **1**: 59-108.

Wagner, M. (2008): The carbon Kuznets curve: A cloudy picture emitted by bad econometrics? Res. Energy Econ. **30**: 388-408.

Waite, F.C. (1901): *Bufo agua* in the Bermudas. Science **13**: 342-343.

Wallach, V. (2008): Range extensions and new island records for *Ramphotyphlops braminus* (Serpentes: Typhlopidae). Bull. Chicago Herpetol. Soc. **43**: 80-82.

Watson, K. (2008): Natural history observations for 2008. J. Barbados Mus. Nat. Hist. Soc. **54**: 261-268.

Weiss, A.J., Hedges, S.B. (2007): Molecular phylogeny and biogeography of the Antillean geckos *Phyllodactylus wirshingi*, *Tarentola americana*, and *Hemidactylus haitianus* (Reptilia, Squamata). Mol. Phylogen. Evol. **45**: 409-416.

Welcomme, R.L. (1988): International Introductions of Inland Aquatic Species. Fish. Tech. Rep. 294. Rome, Food Agricul. Org.

Williams, E.E. (1964): Remarks on the relationship of reptiles and amphibians of the Cayman Islands. Occ. Paps. Molluscs. Mus. Comp. Zool. **2**: 383-384.

Williams, E.E. (1969): The ecology of colonization as seen in the zoogeography of anoline lizards on small islands. Quart. Rev. Biol. **44**: 345-389.

Williams, E.E. (1977): Anoles out of place: Introduced anoles. In: The Third Anolis Newsletter, p. 110-118. Williams, E.E., Ed., Cambridge, Massachusetts, Mus. Comp. Zool.

Williams, E.E., Shreve, B., Humphrey, P.S. (1963): The herpetology of the Port-au-Prince region and Gonave Island, Haiti. Bull. Mus. Comp. Zool. **129**: 293-342.

Williamson, K.E., Poche, A.J., Jr., Greene, B.T., Harris, B.R., Germano, J.M., Simmons, P.M., Yorks, D.T., Powell, R., Parmerlee, J.S., Jr., Henderson, R.W. (2002): Herpetofauna of Hog Island, Grenada. Herpetol. Bull. **82**: 26-29.

Wilson, B.S. (2011): Conservation of Jamaican amphibians and reptiles. In: Conservation of Caribbean Island Herpetofaunas, vol. 2, p. 273-310. Hailey, A., Wilson, B.S., Horrocks, J.A., Eds, Leiden, Brill.

Wilson, B.S., Koenig, S.E., van Veen, R., Miersma, E., Rudolph, D.C. (2010): Cane toads a threat to West Indian wildlife: Mortality of Jamaican boas attributable to toad ingestion. Biol. Invasions **13**: 55-60.

Wilson, S.M. (2001): The prehistory and early history of the Caribbean. In: Biogeography of the West Indies: Patterns and Perspectives, p. 519-527. Woods, C.A., Sergile, F.E., Eds, Boca Raton, Florida, CRC Press.

Wittenberg, R., Cock, M.J.W., Eds (2001): Invasive Alien Species: A Toolkit of Best Prevention and Management Practices. Wallingford, Oxforshire, UK, CAB Intl.

Wolcott, G.N. (1924): The food of Porto Rican lizards. J. Dept. Agricul. P. Rico **7**: 5-37.

Wolcott, G.N. (1934a): The present status of white grub parasites in Puerto Rico. J. Agricul. Univ. P. Rico **18**: 436-441.

Wolcott, G.N. (1934b): The control of white grubs. Intl. Sugar J. **36**: 180.

Wolcott, G.N. (1937): What the giant Suriname Toad, *Bufo marinus* L., is eating now in Puerto Rico. J. Agricul. Univ. P. Rico **21**: 79-84.

Wolcott, G.N. (1948): What has happened to the giant Suriname toad *Bufo marinus* L., in Puerto Rico? Rev. Agricul. P. Rico **38**: 94-98.

Wolcott, G.N. (1950a): The rise and fall of the white grub in Puerto Rico. Amer. Nat. **84**: 183-193.

Wolcott, G.N. (1950b): The sugar-cane rhinoceros beetle. J. Econ. Entomol. **43**: 385.

Wong, M.S., Bundy, D.A.P. (1985): Population distribution of *Ochoterenella digiticauda* (Nematoda: Onchocercidae) and *Mesocoelium monas* (Digenea: Brachycoeliidae) in naturally infected *Bufo marinus* (Amphibia: Bufonidae) from Jamaica. Parasitology **90**: 457-461.

Yanek, K., Heyer, W.R., de Sa, R.O. (2006): Genetic resolution of the enigmatic Lesser Antillean distribution of the frog *Leptodactylus validus* (Anura, Leptodactylidae). S. Amer. J. Herpetol. **1**: 192-201.

Zani, P.A., Guttman, S.I., Powell, R. (1993): The genetic relations of *Anolis cristatellus* (Sauria: Polychridae) from Hispaniola and Puerto Rico. Carib. J. Sci. **29**: 250-253.

Accepted: October 1, 2010 (BSW).

Appendix 1. Species of amphibians and reptiles introduced in the greater Caribbean region. Individual islands within archipelagos (e.g., Bahamas, Virgin Islands) are listed only if introductions from other islands within the same archipelago occurred. Status: W = widespread (likely to be encountered within a few minutes of searching), L = localized (likely to be encountered at most sporadically, even in appropriate habitat, although possibly abundant within a few small areas), E = presumably extirpated or failed introduction, S = stray (no indication of a breeding population ever becoming established). Question marks (?) indicate uncertainty about a published record or, in the case of *Eleutherodactylus johnstonei*, the native range. * = at least some individuals probably introduced intentionally. ** = source almost certainly was populations introduced into Florida or other southeastern US states (although some might be secondary introductions from populations established from Florida stock). Most of the following records are included in the database of introductions in Kraus (2009) and are listed in Schwartz and Henderson (1991) and Henderson and Powell (2009). References cited are those that document or confirm an introduction; all references pertaining to introduced populations are not necessarily listed.

Species (native range)	Introduced (status)	Pertinent reference(s)
FROGS		
Amphibia: Anura: Bufonidae		
Rhinella marina[1]	Anguilla (**S**)	Hodge et al., 2003
(Neotropical mainland)	Antigua* (**W**)	Clark, 1916; Lynn, 1957; Esteal, 1981; Esteal et al., 1981
	Aruba* (**W**)	van Buurt, 2001, 2005, 2006, 2011
	Barbados* (**W**)	Schomburgk, 1848; Gosse, 1851; Waite, 1901; Clark, 1916; Tucker, 1940; Bayley, 1950; Grant, 1959; Esteal, 1981; Esteal et al., 1981; Everard et al., 1988; Forde, 2005; Norville, 2005; Fields and Horrocks, 2009; Horrocks and Fields, 2011
	Canouan (Grenadines) (**S**)	Daudin and de Silva, 2007, 2011
	Carriacou (Grenadines) (**S**)	Lever, 2001, 2003; Daudin and de Silva, 2007, 2011
	Cuba* (**E**)	Bruner, 1935; Jaume, 1966; Buide, 1967; Esteal, 1981; Esteal et al., 1981; Garrido and Jaume, 1984; Estrada and Ruibal, 1999
	Culebra (Puerto Rico) (**?**)	Rivero and Joglar, 1979
	Grand Cayman (**E?**)	Burton and Echternacht, 2003
	Dominica (**E**)	Esteal, 1981; Esteal et al., 1981b; Lever, 2001
	Grenada* (**W**)	Barbour, 1914; Esteal, 1981; Esteal et al., 1981; Everard et al., 1980, 1983; Germano et al., 2003
	Guadeloupe* (**W**)	Jourdane and Theron, 1975; Schwartz and Thomas, 1975; Esteal, 1981; Esteal et al., 1981; Nassi and Dupouy, 1988; Breuil, 2002
	Jamaica* (**W**)	Gosse, 1851; Waite, 1901; Barbour, 1914, 1937; Metcalf, 1914, 1923; Dunn, 1926; Lynn and Grant, 1940; Lynn and Dent, 1943; Goin and Cooper, 1950; Thompson, 1950;

[1] Some insular populations might have become established via natural over-water dispersal.

Species (native range)	Introduced (status)	Pertinent reference(s)
		Mettrick and Dunkley, 1968; Esteal, 1981; Esteal et al., 1981; Crombie et al., 1984; Wong and Bundy, 1985; Wilson et al., 2010; Wilson, 2011
	Hispaniola* (**W**)	Mertens, 1938; Cochran, 1941; Williams et al., 1963; Esteal, 1981; Esteal et al., 1981; Powell et al., 1999
	Martinique* (**W**)	Gosse, 1851; Waite, 1901; Barbour, 1937; Esteal, 1981; Esteal et al., 1981; Breuil, 2009
	Montserrat* (**W**)	Barbour, 1914, 1937; Esteal, 1981; Esteal et al., 1981
	Mustique (Grenadines) (**L**)	Paice, 2005; Daudin and de Silva, 2007, 2011
	Nevis* (**W**)	Barbour, 1914, 1937; Esteal, 1981; Esteal et al., 1981; Horwith and Lindsay, 1999; Lever, 2001
	Puerto Rico* (**W**)	Wolcott, 1924, 1934a, 1934b, 1937, 1948, 1950a, 1950b; Danforth, 1925; May, 1926, 1927, 1930; Grant, 1931; Dexter, 1932; Leonard, 1933; Tucker and Wolcott, 1935; van Volkenberg, 1935; Seín, 1937; Hoffman and Janer, 1941; Pérez, 1951; Cofresi-Sala and Rodríguez de Vega, 1963; Heatwole et al., 1968; Rivero, 1978, 1998; Esteal, 1981; Esteal et al., 1981; Carpenter and Gillingham, 1984, 1987; Rivero and Joglar, 1996; Thomas and Joglar, 1996; Burrowes et al., 2004; Vargas-Salinas, 2005
	St. Christopher* (**W**)	Barbour, 1914, 1937; Esteal, 1981; Esteal et al., 1981; Horwith and Lindsay, 1999
	St. Croix* (**W**)	Grant, 1931; Philbosian and Yntema, 1976; Esteal, 1981; Esteal et al., 1981; MacLean, 1982; Platenberg and Boulon, 2006
	St. Lucia* (**W**)	Barbour, 1914, 1937; Esteal, 1981; Esteal et al., 1981
	St. Vincent* (**W**)	Clark, 1916; Esteal, 1981; Esteal et al., 1981; Censky and Kaiser, 1999; Lever, 2001; Treglia, 2006; Mallery et al., 2007; Powell and Henderson, 2007, 2011
	Union (Grenadines) (**S**)	J. Daudin, pers. comm.
	Virgin Islands* (**W**)	Grant, 1931; McManus and Nellis, 1975; Philbosian and Yntema, 1976; Esteal, 1981; Esteal et al., 1981; MacLean, 1982; Platenberg and Boulon, 2006
Amphibia: Anura: Eleutherodactylidae (formerly assigned to the family Leptodactylidae)		
Eleutherodactylus antillensis (Puerto Rico Bank)	Necker Island (BVI)* (**L**)	Perry and Gerber, 2011
	St. Croix* (**E?**)	Grant, 1937b; Grant and Beatty, 1944; Thomas, 1966
Eleutherodactylus coqui (Puerto Rico)	Bahamas[2] (**?**)	BEST Commission, 2003
	Culebra (**W**)	Rivero and Joglar, 1979; Joglar, 1998

[2] This might very well represent a misidentified animal; the only confirmed introduction in the Bahamas of any species of *Eleutherodactylus* is *E. planirostris*.

Species (native range)	Introduced (status)	Pertinent reference(s)
	Hispaniola* (**L**)	Joglar and Rios-López, 1998
	St. Croix (**L**)	Thomas, 1966; Philbosian and Yntema, 1976, 1977; MacLean, 1982; Platenberg and Boulon, 2006; Waddle et al., 2006
	Vieques (**W**)	Joglar, 1998
	Virgin Islands (**L**)	Thomas, 1966; Philbosian and Yntema, 1976, 1977; MacLean, 1982; Platenberg and Boulon, 2006; Waddle et al., 2006
Eleutherodactylus johnstonei (Antigua Bank?)	Anguilla (**L**)	Censky, 1989; Kaiser and Hardy, 1994; Hodge et al., 2003
	Aruba[3] (**L**)	van Buurt, 2001, 2005, 2006, 2011
	Barbados[4] (**W**)	Feilden, 1889, 1903; Bayley, 1950; Grant, 1959; Schwartz, 1967; Lemon, 1971; Marsh, 1983; Everard et al., 1990; Ovaska, 1991a, 1991b, 1992; Ovaska and Hunte, 1992; Kaiser and Hardy, 1994; Kaiser, 1997; Fields and Horrocks, 2009; Horrocks and Fields, 2011
	Barbuda (**L**)	Kaiser, 1997
	Bequia (Grenadines)*(**L**)	Lazell and Sinclair, 1990; Kaiser and Hardy, 1994; Lazell, 1994; Daudin and de Silva, 2007, 2011
	Bonaire[3] (**L**)	van Buurt, 2001, 2005
	Canouan (Grenadines) (**L**)	Daudin and de Silva, 2007, 2011
	Carriacou (Grenadines) (**L**)	Daudin and de Silva, 2007, 2011
	Curaçao[3] (**W**)	Hardy and Harris, 1979; Kaiser and Hardy, 1994; van Buurt, 2001, 2005, 2006, 2011
	Dominica (**E**[5])	Bullock and Evans, 1990; Corke, 1992; Kaiser, 1992, 1997; Kaiser and Hardy, 1994; Kaiser and Wagenseil, 1995; Daniells et al., 2008
	Grenada (**W**)	Barbour, 1914; Schwartz, 1967; Kaiser and Hardy, 1994; Kaiser and Henderson, 1994; Kaiser, 1997; Goldberg et al., 1998a; Williamson et al., 2002; Germano et al., 2003; Sander et al., 2003; Henderson and Berg, 2005, 2006, 2011
	Guadeloupe (**W**)	Schwartz et al., 1978; Hardy and Harris, 1979; Hardy, 1985; Henderson et al., 1992; Kaiser and Hardy, 1994; Kaiser and Henderson, 1994; Kaiser, 1997; Breuil, 2002; Breuil et al., 2009

[3] Source almost certainly was populations introduced into Venezuela.
[4] Marsh (1983) indicated that this species was native to Barbados.
[5] See Daniells et al. (2008) and Carter et al. (2009).

Species (native range)	Introduced (status)	Pertinent reference(s)
	Jamaica (**W**)	Barbour, 1910; Dunn, 1926; Lynn and Grant, 1940; Perkins, 1942; Lynn and Dent, 1943; Jeffrey-Smith, 1946; Goin and Cooper, 1950; Schwartz and Fowler, 1973; Pough et al., 1977; Stewart, 1977; Stewart and Martin, 1980; Schwartz and Henderson, 1991; Kaiser and Hardy, 1994; Kaiser and Henderson, 1994; Kaiser, 1997; Wilson, 2011
	Les Îles de Saintes (**W**)	Breuil, 2002
	Marie-Galante (**W**)	Henderson et al., 1992; Kaiser and Hardy, 1994; Breuil, 2002; Breuil et al., 2009
	Martinique (**W**)	Lescure, 1966; Kaiser and Henderson, 1994; Lescure and Marty, 1996; Kaiser, 1997; Breuil, 2009; Breuil et al., 2009
	Montserrat (**?**)	Kaiser and Hardy, 1994
	Mustique (Grenadines) (**L**)	Henderson et al., 1992; Kaiser and Hardy, 1994; Daudin and de Silva, 2007, 2011
	Nevis (**?**)	Kaiser and Hardy, 1994; Horwith and Lindsay, 1999
	Petit St. Vincent (Grenadines) (**L**)	Daudin and de Silva, 2007, 2011
	Saba (**W**)	Kaiser and Hardy, 1994; Powell et al., 2005; Powell, 2006
	St.-Barthélemy (**L**)	Kaiser, 1992; Breuil, 2002; Lorvelec et al., 2007; Breuil et al., 2009
	St. Christopher (**?**)	Kaiser and Hardy, 1994; Horwith and Lindsay, 1999
	St. Eustatius (**L**)	Kaiser and Hardy, 1994; Powell et al., 2005; Powell, 2006
	St. Lucia (**L**)	Lescure and Marty, 1996; Lescure, 2000
	St.-Martin/St. Maarten (**L**)	Kaiser and Hardy, 1994; Breuil, 2002; Powell et al., 2005; Powell, 2006
	St. Vincent (**W**)	Lescure, 2000; Treglia, 2006; Mallery et al., 2007; Powell and Henderson, 2007, 2011
	Tortola (BVI) (**L**)	Perry and Gerber, 2011
Eleutherodactylus lentus (USVI)	Jost Van Dyke (BVI)[6] (**W**)	Perry, 2009b; Perry and Gerber, 2011
Eleutherodactylus martinicensis (Antigua, Guadeloupe, Dominica, Martinique)	St.-Barthélemy (**L**)	Kaiser, 1992; Breuil et al., 2009
	St.-Martin/St. Maarten (**L**)	Breuil, 2002; Breuil et al., 2009
Eleutherodactylus planirostris	Great Inagua Island (Bahamas) (**L**)	Schwartz and Henderson, 1991

[6] This species might have been native to the BVI, but an introduction is a more likely explanation (Perry, 2009b).

Species (native range)	Introduced (status)	Pertinent reference(s)
(Cuba, Little and Grand Bahama banks)	Grenada (**S**[7])	Kaiser, 1992; Kraus et al., 1999
	Jamaica (**W**)	Lynn, 1937; Lynn and Grant, 1940; Lynn and Dent, 1943; Goin, 1947; Goin and Cooper, 1950; Schwartz and Fowler, 1973; Schwartz, 1974; Pough et al., 1977; Stewart, 1977; Stewart and Martin, 1980; Wilson, 2011
	Turks & Caicos (**W**)	Schwartz and Henderson, 1991; Reynolds and Niemiller, 2010; Reynolds, 2011
Eleutherodactylus sp. (?)	Guadeloupe (**S**)	Breuil, 2002
	Union Island (Grenadines) (**L**[8])	Henderson et al., 1992; Kaiser and Hardy, 1994; Kaiser and Wagenseil, 1995; Kaiser, 1997; Daudin and de Silva, 2007, 2011
Amphibia: Anura: Hylidae		
Hyla cinerea (Southeastern US)	Puerto Rico (**L**)	Philbosian and Yntema, 1977; Rivero, 1978; Meshaka, 1996; Thomas and Joglar, 1996
Hyla squirella (Southeastern US)	Bahamas (**L**)	Crombie, 1972; Campbell, 1978; Lee, 2004, 2005; Knapp et al., 2011
Osteopilus septentrionalis (Cuba, Bahamas, Cayman Islands)	Anguilla** (**W**)	Townsend et al., 2000; Hodge et al., 2003
	Antigua** (**W**)	Daltry, 2007, 2011
	Bonaire** (**L**)	van Buurt, 2005, 2006, 2011
	Curaçao (**L**)	van Buurt, 2001, 2005, 2006, 2007, 2011
	Dominica** (**S**)	Malhotra et al., 2007, 2011
	Great Inagua (Bahamas) (**W**)	Schwartz, 1968
	Nevis** (**L**)	Lever, 2003; Horwith and Lindsay, 1999
	Puerto Rico** (**W**)	Duellman and Crombie, 1970; Rivero, 1978; Joglar and Rios-López, 1995; Thomas and Joglar, 1996, 1998; Vargas-Salinas, 2006a, 2006b, 2006c
	Saba** (**S**)	Powell, 2006, 2007
	St. Croix** (**W**)	Schwartz and Thomas, 1975; Philbosian and Yntema, 1976, 1977; MacLean, 1982; Waddle et al., 2005; Platenberg and Boulon, 2006

[7] This is almost certainly a spurious record based on an unpublished observation promulgated in the literature.
[8] This frog usually is represented in the literature as *Eleutherodactylus johnstonei*.

Species (native range)	Introduced (status)	Pertinent reference(s)
	St.-Barthélemy** (**W**)	Breuil, 2002; Hodge et al., 2003; Breuil et al., 2009
	St.-Martin/ St. Maarten** (**W**)	Powell et al., 1992[9], 2005; Kaiser and Henderson, 1994; Townsend et al., 2000; Breuil, 2002; Hodge et al., 2003; Breuil et al., 2009; Lorvelec et al., 2011
	Turks & Caicos (**W**)	Reynolds and Niemiller, 2010; Reynolds, 2011
	Virgin Islands** (**W**)	Schwartz and Thomas, 1975; Philbosian and Yntema, 1976, 1977; MacLean, 1982; Meshaka, 1996; Lever, 2003; Owen, 2005; Owen et al., 2005a, 2006; Waddle et al., 2005; Perry and Gerber, 2006, 2011; Perry et al., 2006; Platenberg and Boulon, 2006, 2011; Perry, 2009a
Pseudacris crucifer (Eastern US)	Cuba (**E**[10])	Schwartz and Thomas, 1975; Schwartz and Henderson, 1988, 1991; Estrada and Ruibal, 1999
Scinax ruber (Neotropical mainland)	Martinique (**W**)	Breuil, 2002, 2009; Breuil et al., 2009
	Puerto Rico (**L**)	Thomas and Joglar, 1996; Rivero, 1998; Rios-López, 1999, 2000
	St. Lucia (**L**)	Boulenger, 1891; Barbour, 1914, 1937; Corke, 1992; Kaiser and Henderson, 1994; Censky and Kaiser, 1999
Scinax x-signatus (Neotropical mainland)	Guadeloupe (**W**)	Breuil, 2004; Lorvelec et al., 2007, 2011; Breuil and Ibéné, 2008; Breuil et al., 2009
	Marie-Galante (**L**)	Breuil and Ibéné, 2008; Lorvelec et al., 2011
	Martinique (**L**)	Breuil et al., 2009; Breuil, 2011
Amphibia: Anura: Leiuperidae (formerly assigned to the family Leptodactylidae)		
Pleurodema brachyops (Aruba)	Bonaire, Klein Bonaire (**W**)	Wagenaar Hummelinck, 1940; van Buurt, 2001, 2005
	Curaçao (**W**)	Wagenaar Hummelinck, 1940; van Wijngaarden, 1988; van Buurt, 2001, 2005
Amphibia: Anura: Leptodactylidae		
Leptodactylus fallax (Dominica, Montserrat)	Grenada* (**E**)	Groome, 1970
	Jamaica* (**E**)	Proctor, 1973; Crombie, 1999
	Martinique* (**E**)	Lescure, 1983; Breuil and Ibéne, 2008; Breuil, 2009
	Puerto Rico* (**E**)	May, 1930; Grant, 1931, 1932b; Barbour, 1937; Rivero, 1978; Thomas and Joglar, 1996
Leptodactylus validus (Northern SA)	Bequia (Grenadines) (**L**)	Hardy et al., 2004; Yanek et al., 2006; Camargo et al., 2009
	Grenada (**W**)	Hardy et al., 2004; Yanek et al., 2006; Camargo et al., 2009

[9] Previously listed by Schwartz and Henderson (1991), but misidentified as *Scinax ruber*.

[10] Cuban populations have not been documented and should be removed from lists of West Indian amphibians and reptiles (Powell and Henderson, 1999).

Species (native range)	Introduced (status)	Pertinent reference(s)
	St. Vincent (**W**)	Hardy et al., 2004; Treglia, 2006; Yanek et al., 2006; Camargo et al., 2009
Amphibia: Anura: Microhylidae		
Gastrophryne carolinensis (Southeastern US)	Bahamas (**L**)	Jacobs, 1973a; Crother, 1985; Lee, 2004, 2005; Knapp et al., 2011
	Grand Cayman (**L**)	Schwartz and Henderson, 1991; Seidel and Franz, 1994
Amphibia: Anura: Ranidae		
Lithobates catesbeianus (Eastern US)	Cuba* (**W**)	Hoffman and Noble, 1927; Martínez, 1948; Neill, 1964; Jaume, 1966; Buide, 1967; Odening, 1968; Peters, 1974; Martínez et al., 1982; Coy Otero and Ventosa, 1984; Garrido and Jaume, 1984; Sampedro Marín et al., 1985, 1986, 2003; Coy Otero and Martínez, 1987; de Armas et al., 1987; Novo Rodríguez et al., 1988; Sampedro Marín and Montañez Huguez, 1993; Escobar Herrera, 1995; Montañez et al., 1996; Rueda-Almonacid, 1998, 1999; Estrada and Ruibal, 1999
	Hispaniola* (**W**)	Schwartz and Thomas, 1975; Garrido and Jaume, 1984; Welcomme, 1988; Schwartz and Henderson, 1991; Powell et al., 1999; Neils and Bugbee, 2007
	Jamaica* (**W**)	Grant, 1946; Proctor, 1973; Mahon and Aiken, 1977; Wilson, 2011
	Puerto Rico* (**W**)	Pérez, 1951; Philbosian and Yntema, 1977; Rivero, 1978; Thomas and Joglar, 1996; Joglar, 1998; López-Flores et al., 2003
Lithobates clamitans (Southeastern US)	Bahamas (**S?**)	Lee, 2004, 2005; Knapp et al., 2011
Lithobates grylio (Southeastern US)	Bahamas* (**L**)	Neill, 1964; Schwartz, 1968; Campbell, 1978; Schwartz and Henderson, 1991; Franz et al., 1996; Lee, 2004, 2005; Knapp et al., 2011
	Puerto Rico* (**L**)	Rios-López and Joglar, 1999
Lithobates pipiens (?) (Eastern US)	St. Croix (**E?**)	Grant, 1937
Lithobates sphenocephalus (Southeastern US)	Bahamas (**L**)	Jacobs, 1973b; Lee, 2004, 2005; Knapp et al., 2011
LIZARDS		
Reptilia: Squamata: Anguidae		
Ophisaurus ventralis (Southeastern US)	Grand Cayman (**S**)	Schwartz and Henderson, 1991; Seidel and Franz, 1994

Species (native range)	Introduced (status)	Pertinent reference(s)
Reptilia: Squamata: Corytophanidae		
Basiliscus sp. (Central America)	Bahamas** (**S?**)	Knapp et al., 2011
Reptilia: Squamata: Gekkonidae		
Gekko gecko (Southeastern Asia)	Guadeloupe (**L**)	Breuil, 2004, 2009; Breuil and Ibéné, 2008; Breuil et al., 2010; Lorvelec et al., 2011
	Martinique* (**L**)	Henderson et al., 1993; Breuil, 2009; Breuil et al., 2010
Hemidactylus angulatus (Western Africa)	Cuba (**W**)	Kluge, 1969; Powell and Maxey, 1990; Rodríguez Schettino, 2000; Weiss and Hedges, 2007
	Hispaniola (**W**)	Kluge, 1969; Powell and Maxey, 1990; Powell et al., 1999; Weiss and Hedges, 2007
	Puerto Rico (**W**)	Grant, 1932a; Kluge, 1969; Rivero, 1978; Powell and Maxey, 1990; Weiss and Hedges, 2007
Hemidactylus frenatus (Africa, Southeastern Asia)	Cuba (**L**)	S. Campbell-Staton, pers. comm.
	Hispaniola (**?**)	Scantlebury et al., 2010
Hemidactylus garnotii (Indo-Pacific Region)	Bahamas** (**L**)	Buckner and Franz, 1994a; Meshaka, 1995, 1996; Lee, 2004, 2005; Knapp et al., 2011
Hemidactylus mabouia[11] (Africa, Neotropics)	Aruba (**L**)	Lundberg, 2003; van Buurt, 2005, 2006, 2011
	Bahamas (**L**)	Franz et al., 1993; Buckner and Franz, 1994b; Lee, 2004, 2005; Krysko and Borgia, 2005; Krysko and Thomas, 2007; Knapp et al., 2011
	Bonaire, Klein Bonaire (**L**)	van Buurt, 2001, 2005, 2006, 2011
	Cuba (**L**)	Barbour, 1937; Buide, 1967; Kluge, 1969; Coy Otero and Baruš, 1979; Schwartz and Henderson, 1991; Powell et al., 1998; Estrada and Ruibal, 1999; Martínez Rivera et al., 2003
	Curaçao (**W**)	van Buurt, 2001, 2005, 2006, 2011
	Grand Cayman (**L**)	Echternacht and Burton, 2002

[11] At least some populations in the Lesser Antilles and Virgin Islands might have been established as a consequence of natural over-water dispersal, although such populations might frequently be supplemented by stowaways moving about islands with human mediation; we list only peripheral or recently documented arrivals in this appendix.

Species (native range)	Introduced (status)	Pertinent reference(s)
	Hispaniola (**L**)	Schwartz and Henderson, 1991; Powell et al., 1998, 1999
	Jamaica (**L**)	Fläschendräger, 1999; Wilson, 2011
	Puerto Rico (**W**)	Powell et al., 1998; Mayer, 1999; Thomas, 1999; Martínez Rivera et al., 2003
	Turks & Caicos (**?**)	Minton and Minton, 1975; Reynolds and Niemiller, 2010; Reynolds, 2011
	Virgin Islands (**W**)	Maclean, 1982; Powell et al., 1998; Perry and Gerber, 2011
Hemidactylus palaichthus (Northeastern SA)	Maria Island (off St. Lucia) (**L**)	Kluge, 1969; Powell, 1990c
Hemidactylus turcicus (Mediterranean Region)	Cuba** (**L**)	Leavitt, 1933; Barbour, 1937; Buide, 1967; McCoy, 1970; Baruš and Coy Otero, 1974; Coy Otero and Baruš, 1979; Estrada and Ruibal, 1999; Rodríguez Schettino, 2000
	Puerto Rico** (**S**)	Conant and Collins, 1991
Reptilia: Squamata: Gymnophthalmidae		
Gymnophthalmus underwoodi[12] (Neotropical mainland)	Antigua (**L**)	Powell and Lindsay, 1999; Daltry, 2007, 2011
	Barbados[13] (**W**)	Grant, 1958; Fields and Horrocks, 2009
	Barbuda (**?**)	Censky and Lindsay, 1997
	Bequia (Grenadines) (**?**)	Lazell and Sinclair, 1990
	Dominica (**L**)	Brooks, 1983 (as "*G. pleei*"); Daniells et al., 2008
	Grenada (**L**)	Hardy, 1982
	Guadeloupe (**W**)	Schwartz and Thomas, 1975; Breuil, 2002; Breuil et al., 2010
	Hispaniola (**?**)	Scantlebury et al., 2010
	Marie-Galante (**W**)	Breuil, 2002; Breuil et al., 2010
	Martinique (**L**)	Breuil, 2002, 2009; Breuil et al., 2010
	St. Christopher (**L**)	Orchard, 2010c
	St.-Martin/St Maarten (**L**)	van Buel and Powell, 2006; Breuil, 2009; Breuil et al., 2010; Lorvelec et al., 2011; Powell, 2011
	St. Vincent (**W**)	Schwartz and Thomas, 1975; Treglia, 2006
	Union Island (Grenadines) (**L**)	RP, RWH, pers. obs.

[12] Some insular populations might have become established via natural over-water dispersal.

[13] Fields and Horrocks (2011) implied that the population on Barbados is native.

Species (native range)	Introduced (status)	Pertinent reference(s)
	Virgin Islands (**?**)	Cole et al., 1990
Tretioscincus bifasciatus (Neotropics)	Isla de Providencia (**?**)	Scott and Ayala, 1984; Ayala, 1986; Schwartz and Henderson, 1991; Rueda-Almonacid, 1999
Reptilia: Squamata: Iguanidae		
Ctenosaura similis (Central America)	Bahamas** (**L**)	Knapp et al., 2011
Cyclura cychlura inornata (Leaf and U Cays, Exuma Islands, Bahamas)	Allen Cay (**E**)	Hines and Iverson, 2006a, 2006b
	Cays between Allen and Robert's cays (Exuma Islands)* (**S**)	Hines and Iverson, 2006a, 2006b
Cyclura nubila nubila (Cuba, Lesser Caymans)	Grand Cayman (**E**)	Grant, 1940; Schwartz and Carey, 1977; Seidel and Franz, 1994
	Isla Magueyes (Puerto Rico) (**W**)	Schwartz and Carey, 1977; Rivero, 1978; Christian, 1986; Christian et al., 1986; Christian and Lawrence, 1991; Thomas and Joglar, 1996; Martins and Lamont, 1998; Pérez-Buitrago et al., 2006
	Puerto Rico (L)	M.J. Rivera Rodríguez, pers. comm.
Iguana delicatissima (Lesser Antilles)	Puerto Rico (**S**)	Lever, 2003
Iguana iguana (Neotropics)	Anguilla[14] (**L**)	Censky et al., 1998; Hodge et al., 2003, 2011; Powell, 2004b
	Antigua (**S**)	Powell, 2004b; Powell et al., 2005
	Bahamas** (**L**)	Knapp et al., 2011
	Barbuda (**S**)	Powell, 2004b; Powell et al., 2005
	Grand Cayman (**W**)	Seidel and Franz, 1994; Lever, 2003; Powell and Henderson, 2008
	Guadeloupe[15] (**W**)	Day and Thorpe, 1996; Breuil, 2002; Day et al., 2000; Powell, 2004b; Breuil et al., 2007, 2010
	Les Îles de Saintes[15] (**W**)	Breuil, 2000, 2002; Powell, 2004b; Breuil et al., 2007, 2010

[14] Anguillian populations include released/escaped pets (Hodge et al., 2003) and descendants of animals that arrived via natural rafting (Censky et al., 1998).

[15] The population might or might not be introduced or may consist of descendants of animals that arrived naturally and of others that were introduced.

Species (native range)	Introduced (status)	Pertinent reference(s)
	Marie Galante[15] (**L**)	Breuil, 2002; Powell, 2004b; Breuil et al., 2007, 2010; Lorvelec et al., 2007
	Martinique[15] (**W**)	Breuil, 2000, 2002, 2009; Powell, 2004b; Breuil et al., 2007, 2010; Lorvelec et al., 2007
	Puerto Rico (**W**)	Rivero, 1978; McCoid, 1995; Thomas and Joglar, 1996; Dyer et al., 1999; Thomas, 1999; Engeman et al., 2005; Powell and Henderson, 2008
	St.-Barthélemy (**S**)	Breuil, 2009; Breuil et al., 2010; Lorvelec et al., 2011
	St. Croix[16] (**W**)	Grant, 1937; MacLean, 1982; Thomas and Joglar, 1996; Platenberg and Boulon, 2006
	St.-Martin/St. Maarten (**W**)	Breuil, 2002; Powell, 2004b; Powell et al., 2005; Breuil et al., 2007, 2010; Lorvelec et al., 2007; Powell and Henderson, 2008
	Turks & Caicos** (**S**)	Reynolds and Niemiller, 2010; Reynolds, 2011
	Virgin Islands (**L**)	MacLean, 1982; Thomas and Joglar, 1996; Lazell, 2005; Perry and Gerber, 2006, 2011; Platenberg and Boulon, 2006
Reptilia: Squamata: Polychrotidae		
Anolis bimaculatus (St. Christopher Bank)	Dominica (**S**)	A. James, pers. comm.
	St.-Martin/St. Maarten (**E**)	Powell et al., 1992, 2005
Anolis carolinensis (Southeastern US)	Anguilla (**L**)	Eaton et al., 2001; Hodge et al., 2003
	Grand Bahama (**L**)	Losos et al., 1993
	Grand Cayman (**S**)	Powell, 2002
Anolis cristatellus (Puerto Rico Bank)	Dominica (**L**)	Powell and Henderson, 2003; Malhotra et al., 2007, 2011; Daniells et al., 2008; Ackley et al., 2009
	Dominican Republic (**L**)	Williams, 1969, 1977; Fitch et al., 1989; Schwartz and Henderson, 1991; Zani et al., 1993; Goldberg et al., 1998b; Kolbe et al., 2007a; Powell and Henderson, 2008
	St.-Martin/St. Maarten (**L**)	Breuil et al., 2010
Anolis distichus (Bahamas)	Grand Bahama (**W**)	Dundee, 1990; Schwartz and Henderson, 1991; Losos et al., 1993
	Great Abaco (**L**)	Losos et al., 1993
Anolis equestris (Cuba)	Bahamas** (**L**)	Knapp et al., 2011
	Grand Cayman** (**S**)	Dacosta-Cottam et al., 2010
	Turks & Caicos** (**S**)	Reynolds and Niemiller, 2010; Reynolds, 2011

[16] The current population is almost certainly introduced, but a natural population or animals imported by Amerindians might once have existed.

Species (native range)	Introduced (status)	Pertinent reference(s)
Anolis extremus (Barbados)	St. Lucia (**L**)	Underwood, 1962; Lazell, 1972; Gorman, 1976; Gorman et al., 1978; Corke, 1992; Giannasi et al., 1997
Anolis garmani (Jamaica)	Grand Cayman (**S**)	Schwartz and Henderson, 1991; Seidel and Franz, 1994
Anolis leachii (Antigua Bank)	Virgin Islands (**S**)	Perry, 2005
Anolis lucius (Cuba)	Archipielago de los Canarreos (**L**)	Schwartz and Henderson, 1991
Anolis maynardii (Little Cayman)	Cayman Brac (**L**)	Franz et al., 1987; Seidel and Franz, 1994; Goldberg and Bursey, 1996
Anolis porcatus (Cuba)	Aruba (**L**)	Odum and van Buurt, 2009
	Dominican Republic (**L**)	Arias Cornielle, 1975; Schwartz and Thomas, 1975; Haneline, 1977; Williams, 1977; Vance, 1987; Powell, 1990a, 1990b, 1992; Powell et al., 1990; Powell and Parmerlee, 1991; Parmerlee et al., 1992; Gifford et al., 2002; Kolbe et al., 2007a
Anolis sagrei (Cuba, Bahamas, Lesser Cayman Islands)	Barbados** (**L**)	Fields and Horrocks, 2009, 2011
	Canouan (Grenadines) (**L**)	M. de Silva, pers. comm.
	Grand Cayman** (**W**)	Minton and Minton, 1984; Franz et al., 1987; Lee, 1992; Losos et al., 1993; Seidel and Franz, 1994; Goldberg et al., 1995; Gerber and Echternacht, 2000; Kolbe et al., 2004, 2007a, 2007b
	Grenada** (**L**)	Greene et al., 2002; Germano et al., 2003; Kolbe et al., 2004
	Jamaica (**W**)	Gosse, 1850; Underwood and Williams, 1959; Williams, 1969; Schoener and Schoener, 1971; Landwer et al., 1995; Bundy et al., 1987; Landwer and Ferguson, 2002; Kolbe et al., 2004; Wilson, 2011
	St.-Martin/ St. Maarten** (**L**)	Fläschendräger, 2010
	St. Vincent** (**L**)	Henderson and Powell, 2005; Treglia, 2006; Mallery et al., 2007; Powell and Henderson, 2007, 2011; Treglia et al., 2008
Anolis wattsi (Antigua)	St. Lucia (**L**)	Underwood, 1959, 1962; Lazell, 1972; Gorman, 1976; Corke, 1992

Species (native range)	Introduced (status)	Pertinent reference(s)
Reptilia: Squamata: Sphaerodactylidae		
Gonatodes albogularis	Aruba (**E**)	Wagenaar Hummelinck, 1940; van Buurt, 2001, 2005
(Neotropics, Cuba?)	Grand Cayman (**L**)	Williams, 1964; Seidel and Franz, 1994
	Curaçao (**E**)	Wagenaar Hummelinck, 1940; van Buurt, 2001, 2005
	Hispaniola (**L**)	Crombie, 1999
	Jamaica (**L**)	Crombie, 1999; Wilson, 2011
Gonatodes antillensis (Bonaire, Curaçao)	Aruba (**E**)	Odum, 1992; van Buurt, 2001, 2005
Gonatodes vittatus	Aruba (**W**)	Wagenaar Hummelinck, 1940; Lundberg, 2003; van Buurt, 2001, 2005
(Neotropics, Aruba?)	Curaçao (**L**)	Wagenaar Hummelinck, 1940; van Buurt, unpubl. data
	Dominica (**E**)	Malhotra et al., 2007, 2011
Sphaerodactylus argus	Bahamas (**W**)	Barbour, 1937; Schwartz, 1968; Thomas, 1975
(Cuba, Jamaica)	Cuba (**L**)	Barbour, 1937; Savage, 1954; Buide, 1967; Thomas, 1975; Estrada and Ruibal, 1999
Sphaerodactylus copei (Hispaniola)	Bahamas** (**L**)	Schwartz, 1968; Franz et al., 1996; Lee, 2004, 2005; Knapp et al., 2011
Sphaerodactylus mariguanae (Mayaguana and Booby Cay)	Grand Turk (Turks & Caicos) (**E?**)	Schwartz and Henderson, 1991; Reynolds and Niemiller, 2010; Reynolds, 2011
Sphaerodactylus microlepis (St. Lucia)	Dominica (**S**)	Evans, 1989; Malhotra and Thorpe, 1999
Sphaerodactylus notatus	Great Inagua (Bahamas) (**?**)	Schwartz, 1965, 1970, 1973
(Bahamas, Cuba?)	Morant Cays (Jamaica) (**?**)	Schwartz, 1965, 1970, 1973
Reptilia: Squamata: Teiidae		
Ameiva ameiva (Trinidad?)	Barbados (**L**)	Corrie, 2001; Watson, 2008; Fields and Horrocks, 2009, 2011
Ameiva exsul (Virgin Islands)	St. Croix (**L**)	Platenberg and Boulon, 2006

Species (native range)	Introduced (status)	Pertinent reference(s)
Cnemidophorus lemniscatus (Neotropics)	Aruba (**L**)	Schall, 1973; van Buurt, 2001, 2005
Tupinambis teguixin (Neotropics)	Isla de San Andres (**?**)	Rueda-Almonacid, 1999
Reptilia: Squamata: Varanidae		
Varanus exanthematicus (Africa)	Turks & Caicos (**S**)	Reynolds, 2011
AMPHISBAENIANS		
Reptilia: Squamata: Amphisbaenidae		
Amphisbaena fuliginosa (SA and Trinidad)	Grenada (**S**?)	Murphy et al., 2010
	St. Lucia (**S**?)	Murphy et al., 2010
SNAKES		
Reptilia: Squamata: Boidae		
Boa constrictor (Neotropics)	Aruba (**L**)	Quick et al., 2005; van Buurt, 2001, 2005, 2006, 2011
	Bonaire (**S**)	van Buurt, 2001, 2005
	Curaçao (**S**)	van Buurt, 2001, 2005, 2006, 2011
	Guadeloupe*[17] (**S**)	Barré et al., 1997; Breuil, 2002; Breuil et al., 2010
	Martinique (**S**)	Breuil, 2009; Breuil et al., 2010
	Puerto Rico (**L**)	Krysko and King, 2010; USFWS, 2010; M.J. Rivera Rodríguez and A.J. Sánchez Muñoz, pers. comm.; R. Reed and G.H. Rodda, pers. comm.
	St.-Martin/St. Maarten (**S**)	Powell et al., 2005
Epicrates cenchria (Neotropics)	Martinique (**S**)	Breuil and Ibéné, 2008; Breuil, 2009; Breuil et al., 2010
	St.-Martin/St. Maarten (**S**)	Powell et al., 2005
Eunectes notaeus (SA mainland)	Puerto Rico (**S**)	USFWS, 2010; R. Reed and G.H. Rodda, pers. comm.
Reptilia: Squamata: Colubridae		
Leptophis sp. (SA mainland)	Curaçao (**S**)	van Buurt, 2001, 2005

[17] This record might pertain to *Boa nebulosa* (Lorvelec et al., 2011).

Species (native range)	Introduced (status)	Pertinent reference(s)
Mastigodryas bruesi (Grenada or St. Vincent bank)	Barbados (**L**)	Underwood et al., 1999; Greene et al., 2003; Powell and Henderson, 2007, 2011; Fields and Horrocks, 2009, 2011
Opheodrys aestivus (Eastern US)	Bahamas (**L**)	Knapp et al., 2011
Pantherophis alleghaniensis[18] (Eastern US)	Bahamas (**L**)	Buckner and Franz, 1994d; Lee, 2004, 2005; Knapp et al., 2011
Pantherophis guttatus (Southeastern US)	Anguilla (**S**)	Hodge et al., 2003
	Antigua (**S**)	Powell and Henderson, 2003
	Bahamas (**L**)	Buckner and Franz, 1994c; Lee, 2004, 2005; Knapp et al., 2011
	Bonaire (**S**)	Perry et al., 2003; van Buurt, 2001, 2005, 2006, 2011
	Curaçao (**E?**)	Perry et al., 2003; van Buurt, 2001, 2005, 2006, 2011
	Grand Cayman (**L**)	Franz et al., 1987; Schwartz and Henderson, 1991; Seidel and Franz, 1994; Tolson and Henderson, 2011
	Martinique (**S**)	Breuil, 2009; Breuil et al., 2010
	St.-Barthélemy (**S**)	Breuil, 2002; Hodge et al., 2003; Breuil et al., 2010
	St.-Martin/St. Maarten (**S**)	Powell et al., 2005; Breuil et al., 2010
	Turks & Caicos (**S**)	Reynolds and Niemiller, 2010; Reynolds, 2011; Tolson and Henderson, 2011
	Virgin Islands (**L**)	Hodge et al., 2003; Perry et al., 2003, 2011; Platenberg and Boulon, 2006; Tolson and Henderson, 2011
Tantilla melanocephala (Trinidad, SA)	Carriacou (Grenadines) (**S**)	D. Scantlebury and J. Boone, pers. comm.
	Grenada (**L**)	Berg et al., 2009; Tolson and Henderson, 2011
	Mustique (Grenadines)	Henderson and Powell, 2006; Berg et al., 2009; Tolson and Henderson, 2011
	Union (Grenadines)	Berg et al., 2009; Tolson and Henderson, 2011
Reptilia: Squamata: Dipsadidae		
Alsophis rufiventris (Saba, St. Eustatius)	St.-Martin/St. Maarten (**S**)	Powell et al., 2005

[18] This species also has been assigned to the genus *Scotophis*.

Species (native range)	Introduced (status)	Pertinent reference(s)
Borikenophis portoricensis	St. Croix (**S**)	Perry and Platenberg, 2007
(Puerto Rico Bank)	St. Thomas (USVI)[19] (**W**)	Platenberg and Boulon, 2011
Diadophis punctatus	Curaçao (**S**)	van Buurt, 2001, 2005
(NA mainland)	Grand Cayman (**S**)	Schwartz and Henderson, 1991; Seidel and Franz, 1994
Hypsirhynchus parvifrons (Hispaniola)	Little Inagua (Bahamas) (**S**)	Schwartz and Thomas, 1975
Imantodes sp. (SA mainland)	Curaçao (**S**)	van Buurt, 2001, 2005
Leptodeira sp. (SA mainland)	Curaçao (**S**)	van Buurt, 2001, 2005
Reptilia: Squamata: Elapidae		
Micrurus fulvius (Southeastern US)	Curaçao (**S**)	van Buurt, 2001, 2005
Reptilia: Squamata: Leptotyphlopidae		
Epictia albifrons (SA)	Bonaire (**L**)	Wagenaar Hummelinck, 1940; van Buurt, 2001, 2005, 2006, 2011
Reptilia: Squamata: Natricidae		
Natrix natrix (Europe)	Martinique (**S**)	Breuil and Ibéné, 2008; Breuil, 2009; Breuil et al., 2010; Lorvelec et al., 2011
Storeria dekayi (Eastern US)	Bahamas (**L**)	Lee, 2004, 2005; Knapp et al., 2011; Tolson and Henderson, 2011
Thamnophis cyrtopsis (Southwestern NA)	Curaçao (**S**)	van Buurt, 2001, 2005
Thamnophis sauritus (Eastern US)	Bahamas (**S**)	Buckner and Franz, 1998a; Lee, 2004, 2005; Knapp et al., 2011

[19] This was identified as a "recolonization" (Platenbergy and Boulon, 2011).

Species (native range)	Introduced (status)	Pertinent reference(s)
Thamnophis sirtalis	Bahamas (**S**)	Buckner and Franz, 1998b; Lee, 2004, 2005; Knapp et al., 2011
(Eastern US)	St.-Barthélemy (**S**)	Breuil et al., 2010
Reptilia: Squamata: Pythonidae		
Morelia amethistina (Indonesia, Papua New Guinea, Australia)	Guadeloupe (**S**)	Breuil and Ibéné, 2008; Breuil et al., 2010; Lorvelec et al., 2011
Python bivittatus	Puerto Rico (**S**)	ISSG, 2010; USFWS, 2010; R. Reed and G.H. Rodda, pers. comm.
(Southeastern Asia)	St.-Barthélemy (**S**)	Breuil et al., 2010
Python curtus group (Malaya, Indonesia)	St.-Martin/St. Maarten (**S**)	Powell et al., 2005
Python regius	Martinique (**S**)	Breuil, 2009; Breuil et al., 2010
(West-central Africa)	Puerto Rico (**S**)	USFWS, 2010; R. Reed and G.H. Rodda, pers. comm.
	St.-Martin/St. Maarten (**S**)	Powell et al., 2005
	St.-Barthélemy (**S**)	Breuil et al., 2010
Python reticulatus (Southeastern Asia)	Puerto Rico (**S**)	USFWS, 2010; R. Reed and G.H. Rodda, pers. comm.
Python sebae (Subsaharan Africa)	Puerto Rico (**S**)	USFWS, 2010; R. Reed and G.H. Rodda, pers. comm.
Reptilia: Squamata: Typhlopidae		
Ramphotyphlops braminus	Anguilla** (**L**)	Censky and Hodge, 1997; Hodge et al., 2003
(Southeastern Asia)	Aruba** (**L**)	van Buurt, 2006, 2011
	Barbados** (**W**)	Hedges, 2008; Fields and Horrocks, 2009, 2011
	Curaçao** (**L**)	Wallach, 2008
	Grand Cayman** (**L**)	Echternacht and Burton, 2003; Hodge et al., 2003; Tolson and Henderson, 2011
	Guadeloupe** (**L**)	Breuil and Ibéné, 2008; Breuil, 2009; Breuil et al., 2010; Lorvelec et al., 2011
	Mustique (Grenadines)** (**?**)	M. de Silva, in litt., 2009
	St.-Barthélemy** (**L**)	Breuil, 2002; Hodge et al., 2003; Breuil et al., 2010
	St. Christopher** (**?**)	Orchard, 2010a

Species (native range)	Introduced (status)	Pertinent reference(s)
	St. Eustatius** (**L**)	Powell, 2011
	St.-Martin/St. Maarten** (**L**)	Breuil, 2002; Hodge et al., 2003; Powell et al., 2005; Breuil et al., 2010
	Turks & Caicos** (**W**)	Reynolds and Niemiller, 2010; Reynolds, 2011; Tolson and Henderson, 2011
TURTLES		
Reptilia: Testudines: Chelidae		
Phrynops geoffroanus (SA)	Anguilla (**S**)	Hodge et al., 2011
Reptilia: Testudines: Emydidae		
Graptemys pseudogeographica (Central US)	Martinique (**S**)	Breuil, 2009; Breuil et al., 2010
Pseudemys nelsoni (Florida)	Virgin Islands (**S**)	Owen et al., 2005b; Perry and Gerber, 2006
Terrapene carolina (Eastern US)	Bahamas (**S?**)	Lee, 2004, 2005; Knapp et al., 2011
	Martinique (**S**)	Breuil, 2009; Breuil et al., 2010
Trachemys decorata (Hispaniola)	Bahamas (**L**)	Lee, 2004, 2005; Knapp et al., 2011
Trachemys decussata (Cuba)	Grand Cayman (**L**)	Dunson and Seidel, 1986; Alderton, 1988; Seidel, 1988, 1990, 1996, 2003; Seidel and Franz, 1994
Trachemys scripta (Eastern US)	Aruba (**L**)	van Buurt, 2005
	Bahamas (**L**)	Lee and Carey, 2001; Lee and Ross, 2001; Mealey et al., 2002; Lee, 2004, 2005; Knapp et al., 2011
	Barbados (**S**)	Fields and Horrocks, 2011
	Grand Cayman[20] (**S**)	Lever, 2003
	Hispaniola (**W**)	Powell et al., 2000; Powell and Incháustegui, 2009, 2011
	Guadeloupe (**S**)	Schwartz and Thomas, 1975; Lescure, 1979; Schwartz and Henderson, 1988, 1991; Breuil, 2002; Breuil et al., 2010

[20] Possibly a misidentified *Trachemys decussata*.

Species (native range)	Introduced (status)	Pertinent reference(s)
	Marie-Galante (**S**)	Breuil, 2002
	Martinique (**S**)	Servan and Arvy, 1997; Breuil, 2002
	St. Croix (**L**)	Platenberg and Boulon, 2006
	St. Eustatius (**S**)	Powell et al., 2005
	St.-Martin/St. Maarten (**W**)	Powell et al., 2005
	Turks & Caicos (**L**)	Reynolds and Niemiller, 2010; Reynolds, 2011
	Virgin Islands (**L**)	Owen et al., 2005b; Perry and Gerber, 2006, 2011; Platenberg and Boulon, 2006; Perry et al., 2007
Trachemys stejnegeri (Turks & Caicos,	Bahamas (**L**)	Hodsdon and Pearson, 1943; Campbell, 1978; Groombridge, 1982; Seidel and Adkins, 1987; Seidel, 1988; Lee and Ross, 2001; Knapp et al., 2011
Hispaniola, Puerto Rico)	Culebra (Puerto Rico) (**S**)	Seidel, 1988
	Dominica (**E**)	Fritz, 1991; Seidel, 1996
	Guadeloupe (**W**)	Breuil, 2002, 2003; Breuil et al., 2010
	Les Îles de Saintes* (**L**)	Breuil, 2002
	Marie-Galante (**L**)	Seidel and Adkins, 1987; Seidel, 1988; Ernst and Barbour, 1989; Breuil, 2002, 2003; Breuil et al., 2010
	Vieques (Puerto Rico) (**?**)	Seidel, 1988
	Turks & Caicos (**E?**)	Reynolds and Niemiller, 2010; Reynolds, 2011
Trachemys terrapen (Bahamas, Jamaica?)	New Providence (Bahamas) (**L**)	Campbell, 1978: Pritchard, 1979; Groombridge, 1982; Ross, 1982; Seidel and Adkins, 1987; Seidel, 1988, 1996; Lee and Ross, 2001; Knapp et al., 2011
Trachemys sp. (?)	Bahamas (**L**)	Franz et al., 1993; Seidel, 1996; Knapp et al., 2011
Reptilia: Testudines: Pelomedusidae		
*Pelusios castaneus**[21] (Western Africa)	Guadeloupe (**W**)	Lescure, 1979, 1983; Schwartz and Henderson, 1991; Iverson, 1992; Breuil, 2002, 2003; Breuil et al., 2010
Reptilia: Testudines: Testudinidae		
Centrochelys sulcata	Necker & Guana islands (BVI) (**S**)	G. Perry, unpubl. data

[21] Previously misidentified as *Pelusios subniger*.

Species (native range)	Introduced (status)	Pertinent reference(s)
(Northern Africa)	Martinique (**S**)	Breuil, 2009; Breuil et al., 2010
Chelonoidis carbonaria (Neotropics)	Barbados (**L**)	Fields and Horrocks, 2009, 2011
	Saba (**S**)	Powell et al., 2005
	St.-Barthélemy* (**L**)	Breuil, 2004
	St. Christopher (**L**)	Horwith and Lindsay, 1999; Orchard, 2010b
	St. Eustatius* (**S**)	Powell et al., 2005
Chelonoidis denticulata (Neotropical mainland)	Guadeloupe (**L**)	Pritchard and Trebbau, 1984; Breuil, 2002; Breuil et al., 2010
*Kinixys erosa** (Western Africa)	Guadeloupe (**E**)	Breuil, 2002, 2003
*Kinixys homeana** (Western Africa)	Guadeloupe (**E**)	Breuil, 2002, 2003
CROCODILIANS		
Reptilia: Crocodilia: Alligatoridae		
Alligator mississippiensis (Southeastern NA)	Bahamas (**S**)	Carey, 2002; Lee, 2004, 2005; Knapp et al., 2011
Caiman crocodilus (Neotropics)	Carriacou (Grenadines) (**S**)	Devas, 1964; Groome, 1970
	Isla de la Juventud (Cuba)* (**L**)	Varona, 1976, 1980, 1981; Groombridge, 1982; Garrido and Jaume, 1984; Escobar Herrera, 1995; Estrada and Ruibal, 1999
	Isla de San Andres (**S**)	Rueda-Almonacid, 1999
	Puerto Rico (**S**)	Schwartz and Henderson, 1985, 1991; Thomas and Joglar, 1996; Thomas, 1999
	Vieques (Puerto Rico) (**S**)	Thomas, 1999
Reptilia: Crocodilia: Crocodylidae		
Crocodylus acutus? (Neotropics)	Klein Curaçao (**S**)	van Buurt, 2001, 2005
Crocodylus intermedius (SA)	Grenada (**S**)	Groome, 1970

Appendix 2. Species of amphibians and reptiles introduced in the greater Caribbean region for research or conservation (including restorations). Individual islands within archipelagos (e.g., Bahamas, Virgin Islands) are listed only if introductions from other islands within the same archipelago occurred. Status: W = widespread (likely to be encountered within a few minutes of searching), L = localized (likely to be encountered at most sporadically, even in appropriate habitat, although possibly abundant within a few small areas), E = presumably extirpated or failed introduction. Most of the following records are included in the database of introductions in Kraus (2009) and listed in Henderson and Powell (2009). References cited are those that document or confirm an introduction; all references pertaining to introduced populations are not necessarily listed.

Species (native range)	Introduced (status)	Pertinent reference(s)
FROGS		
Amphibia: Anura: Eleutherodactylidae (formerly assigned to the family Leptodactylidae)		
Eleutherodactylus cochranae (Puerto Rico)	Puerto Rico (Isla Palominitos) (**E**)	Levins and Heatwole, 1973
Eleutherodactylus schwartzi (Great Dog Island, BVI)	Little Thatch Island (BVI) (**L**)	Lazell, 2002, 2005; Perry and Gerber, 2006
LIZARDS		
Reptilia: Squamata: Iguanidae		
Cyclura carinata (Turks & Caicos)	Cays in the Turks & Caicos[22] (**L**)	G. Gerber, in litt., 2010
Cyclura cychlura figginsi (Exuma Islands)	Pasture Cay (**L**)	Knapp, 2002
Cyclura cychlura inornata (Leaf and U Cays, Exuma Islands, Bahamas)	Alligator Cay (**L**)	Iverson, 2000; Knapp and Malone, 2003; Iverson et al., 2006
	Flat Rock Reef Cay (**L**)	Hines and Iverson, 2006a, 2006b
Cyclura pinguis (Anegada Island, BVI)	Guana Island (BVI) (**W**)	Kirby, 1986; Goodyear and Lazell, 1994; Lazell, 2002, 2005; Perry and Gerber, 2006, 2011
(Necker Island)	Moskito (BVI) (**L**)	Perry and Gerber, 2011
(Guana Island)	Necker Island (BVI) (**W**)	Lazell, 1995, 2000, 2002, 2005, 2006; Perry and Gerber, 2006, 2011

[22] These efforts involve relocation of animals from an island or islands that are to be intensely developed to islands that currently lack iguanas, although whether or not they supported populations in the past is unknown.

Species (native range)	Introduced (status)	Pertinent reference(s)
Cyclura rileyi nuchalis (Acklin Bight, Bahamas)	Bush Hill Cay (**L**)	Hayes et al., 2004
Iguana delicatissima (Îlet Chancel, Martinique)	Îlet à Ramiers (Martinique) (**L**)	Breuil, 2009
Reptilia: Squamata: Leiocephalidae		
Leiocephalus carinatus (Bahamas)	Bahamian Cays[23] (**E**)	Schoener and Spiller, 1996; Spiller et al., 1998; Schoener et al., 2001, 2002, 2005
Reptilia: Squamata: Polychrotidae		
Anolis lineatus (Curaçao)	Klein Curaçao (**L**)	van Buurt, 2006, 2011
Anolis pogus (Anguilla Bank)	Anguillita (**E**)	Roughgarden et al., 1984
Anolis pulchellus (Puerto Rico)	Isla Palominitos (**E**)	Levins and Heatwole, 1973
Anolis sagrei (Bahamas)	Bahamian Cays[24] (**L**)	Calsbeek and Smith, 2007; Calsbeek et al., 2008; Calsbeek, 2009
Anolis stratulus (Puerto Rico)	Isla Palominitos (**E**)	Levins and Heatwole, 1973
Reptilia: Squamata: Teiidae		
Ameiva polops (St. Croix)	Buck Island (St. Croix) (**E**)	Philibosian and Ruibal, 1971; Philibosian and Yntema, 1976; Dodd, 1978; Platenberg and Boulon, 2011
	Ruth Island (St. Croix) (**L**)	Knowles, 1990, 1997; McNair and Mackay, 2005
Cnemidophorus vanzoi (Maria Major, St. Lucia)	Praslin Island (**W**)	Dickinson and Fa, 2000

[23] Small cays with *Anolis sagrei* present, but previously without *L. carinatus*.

[24] Lizards (*A. sagrei*) were removed and/or translocated to manipulate population densities.

Species (native range)	Introduced (status)	Pertinent reference(s)
SNAKES		
Reptilia: Squamata: Dipsadidae		
Alsophis sajdaki (Great Bird Island, Antigua)	Green and Rabbit islands (**L**)	Daltry, 2007, 2011
TURTLES		
Reptilia: Testudines: Testudinidae		
Chelonoidis carbonaria (Tortola and Water Island)	Guana Island (BVI) (**W**)	Lazell, 2002, 2005; Perry and Gerber, 2006
	Necker Island (BVI) (**L**)	Lazell, 2002, 2005, 2006; Perry and Gerber, 2006

Appendix 3. Species exported from the US to the countries indicated. Those marked with an asterisk (*) do not occur (native or introduced) in the country to which it is being exported (USFWS LEMIS database). Those marked with a double-asterisk (**) do not occur in the greater Caribbean. A&B = Antigua and Barbuda, CI = Cayman Islands, DR = Dominican Republic, NA = Netherlands Antilles, PR = Puerto Rico, T&C = Turks & Caicos Islands, USVI = US Virgin Islands.

Species	Country	1998	1999	2000	2001	2002	2003	2004	2005	2006	2007	2008	2009	Total
Peltophryne (= *Bufo*) *lemur*	PR	–	–	–	–	–	–	–	–	–	700	–	–	700
*Bombina orientalis***	DR	–	3612	–	–	–	–	–	–	–	–	–	–	3612
	NA	–	–	–	–	150	–	–	–	–	–	–	–	150
Bombina sp.**	DR	10	–	–	–	–	–	–	–	–	–	–	–	10
*Cryptophyllobates azureiventris***	NA	–	–	–	–	–	–	–	–	–	–	30	–	30
*Dendrobates auratus***	NA	–	–	–	–	–	–	–	–	–	–	15	–	15
*Dendrobates azureus***	NA	–	–	–	–	–	–	–	–	–	–	105	–	105
*Dendrobates imitator***	NA	–	–	–	–	–	–	–	–	–	–	6	–	6
*Dendrobates tinctorius***	NA	–	–	–	–	–	–	–	–	–	–	80	–	80
*Dendrobates ventrimaculatus***	NA	–	–	–	–	–	–	–	–	–	–	20	–	20
*Phyllobates terribilis***	NA	–	–	–	–	–	–	–	–	–	–	15	–	15
*Hyla cinerea**	NA	12	–	–	–	25	–	–	–	–	–	–	–	37
Hyla sp.*	PR	–	–	–	–	–	–	–	1	–	–	–	–	1
*Ceratophrys cranwelli***	NA	–	–	–	–	25	–	–	–	–	–	–	–	25

Species	Country	1998	1999	2000	2001	2002	2003	2004	2005	2006	2007	2008	2009	Total
*Ceratophrys ornata***	Jamaica	45	–	–	–	–	–	–	–	–	–	–	–	45
*Hymenochirus curtipes***	Bahamas	–	15	6	–	–	–	–	–	–	–	–	–	21
	Barbados	–	–	–	–	–	–	–	50	–	–	–	–	50
	CI	–	–	–	–	–	12	–	–	–	–	–	–	12
*Xenopus laevis***	Bahamas	–	45	–	–	–	–	–	–	–	–	–	12	57
	Barbados	–	–	–	–	200	–	–	–	–	–	–	–	200
	DR	–	–	–	20	–	–	–	–	–	–	–	–	20
Lithobates catesbeianus (= *Rana catesbeiana*)	DR	4900	–	–	–	–	–	–	–	–	–	–	–	4900
Ambystoma sp.**	Bahamas	–	–	–	24	–	–	–	–	–	–	–	–	24
	CI	–	6	12	–	–	–	–	–	–	–	–	–	18
*Ambystoma tigrinum***	Bahamas	–	–	10	–	–	–	13	–	–	–	–	–	23
	CI	–	–	6	12	20	–	–	–	–	–	–	–	38
	DR	–	–	–	22	–	–	–	–	–	–	–	–	22
*Necturus maculosus***	Barbados	–	–	–	–	–	–	–	20	–	–	–	–	20
*Cynops pyrrhogaster***	Bahamas	–	35	16	–	12	12	12	–	6	–	–	–	93
	Barbados	–	–	–	–	–	30	–	–	–	20	–	–	50
	CI	–	12	1012	24	73	84	–	–	–	–	–	–	1205
Cynops sp.**	Bahamas	–	–	–	–	–	–	12	–	6	12	–	–	30
	Barbados	–	–	–	–	–	–	–	–	40	–	–	–	40
Notophthalmus sp.**	Bahamas	–	–	–	–	–	–	–	–	12	–	–	–	12
	CI	–	–	–	–	–	12	–	–	–	–	–	–	12
*Notophthalmus viridescens***	CI	–	–	–	12	12	12	–	–	–	–	–	–	36
*Paramesotriton hongkongensis***	Barbados	–	–	–	–	–	–	–	–	–	100	–	–	100
Taricha sp.**	CI	–	–	–	12	–	–	–	–	–	–	–	–	12
*Taricha torosa***	Bahamas	–	–	–	–	–	12	–	–	6	–	–	–	18
	Barbados	–	–	–	–	–	–	–	–	20	–	–	–	20
	CI	–	–	511	–	24	–	–	–	–	–	–	–	535
	DR	–	–	–	–	–	–	–	–	40	–	–	–	40
Non-CITES entry (amphibian)	Bahamas	–	–	–	–	–	–	–	–	56	48	84	–	188
	DR	35	–	–	–	–	–	–	–	–	–	–	–	35

Species	Country	1998	1999	2000	2001	2002	2003	2004	2005	2006	2007	2008	2009	Total
Alligator mississippiensis	Bahamas	12	–	–	–	–	–	–	–	–	–	–	–	12
*Furcifer pardalis***	Dominica	–	–	–	–	–	–	6	–	–	–	–	–	6
*Coleonyx mitratus***	NA	–	–	–	–	6	–	–	–	–	–	–	–	6
*Eublepharis macularius***	Barbados	–	–	–	–	–	–	–	12	–	–	–	–	12
*Gekko gecko**	NA	–	–	–	–	12	–	–	–	–	–	–	–	12
Gekko sp.**	Barbados	–	–	–	–	–	1	–	24	–	–	–	–	25
Hemidactylus sp.	NA	–	–	–	–	100	–	–	–	–	–	–	–	100
Ptychozoon sp.**	NA	–	–	–	–	12	–	–	–	–	–	–	–	12
*Basiliscus plumifrons***	NA	1	–	–	–	–	–	–	–	–	–	–	–	1
*Basiliscus vittatus***	NA	–	–	–	–	12	–	–	–	–	–	–	–	12
*Corytophanes cristatus***	NA	2	–	–	–	–	–	–	–	–	–	–	–	2
Cyclura nubila	CI	–	2	–	–	–	–	–	–	–	–	–	–	2
*Iguana iguana**	Bahamas	–	–	–	–	–	4	–	–	–	–	–	–	4
Iguana iguana	NA	–	–	–	–	25	–	–	–	45	20	–	–	90
	PR	–	–	1	–	–	–	–	–	–	–	–	–	1
*Leiocephalus carinatus**	Haiti	–	–	250	–	–	–	–	–	–	–	–	–	250
Leiocephalus sp.	Haiti	–	–	200	–	–	–	–	–	–	–	–	–	200
*Sceloporus malachiticus***	NA	–	–	–	–	12	–	–	–	–	–	–	–	12
*Sceloporus olivaceus***	NA	–	–	–	–	4	–	–	–	–	–	–	–	4
*Sceloporus variabilis***	NA	–	–	–	–	12	–	–	–	–	–	–	–	12
Takydromus sp.**	NA	–	–	–	–	100	–	–	–	–	–	–	–	100
*Anolis equestris**	NA	2	–	–	–	–	–	–	–	–	–	–	–	2
*Anolis carolinensis**	NA	–	–	–	–	100	–	–	–	–	–	–	–	100
Anolis sagrei	NA	–	–	–	–	100	–	–	–	–	–	–	–	100
Anolis sp.	NA	–	–	–	–	6	–	–	–	–	–	–	–	6
Novoeumeces (= *Eumeces*) *schneideri***	NA	–	–	–	–	12	–	–	–	–	–	–	–	12
*Ameiva ameiva**	NA	–	–	–	–	12	–	–	–	–	–	–	–	12
Cnemidophorus lemniscatus	NA	–	–	–	–	12	–	–	–	–	–	–	–	12
*Boa constrictor**	Dominica	–	–	–	–	–	8	–	–	–	–	–	–	8
	DR	–	1	–	–	–	–	–	–	–	–	–	–	1
	NA	–	–	–	–	–	–	–	–	5	–	–	–	5

Species	Country	1998	1999	2000	2001	2002	2003	2004	2005	2006	2007	2008	2009	Total
*Corallus caninus***	Barbados	–	–	–	3	–	–	–	–	–	–	–	–	3
*Corallus hortulanus***	Barbados	–	–	–	1	–	–	–	–	–	–	–	–	1
*Epicrates cenchria**	A&B	2	–	–	–	–	–	–	–	–	–	–	–	2
	Barbados	–	–	–	2	–	–	–	–	–	–	–	–	2
*Eunectes murinus***	Barbados	–	–	–	1	–	–	–	–	–	–	–	–	1
Pantherophis guttatus (= *Elaphe guttata*)*	Barbados	–	–	–	–	–	–	–	6	–	–	–	–	6
Pantherophis guttatus (= *Elaphe guttata*)	NA	–	–	–	–	6	–	–	–	–	–	–	–	6
*Pantherophis obsoletus** (= *Elaphe obsoleta*)	Barbados	–	–	–	–	–	–	–	3	–	–	–	–	3
*Elaphe quadrivirgata***	Barbados	–	–	–	–	–	–	–	1	–	–	–	–	1
Pantherophis vulpinus (= *Elaphe vulpina*)**	NA	–	–	–	–	11	–	–	–	–	–	–	–	11
*Lampropeltis calligaster***	Barbados	–	–	–	–	–	–	–	1	–	–	–	–	1
*Lampropeltis getula***	NA	–	–	–	–	–	–	–	–	–	4	–	–	4
Lampropeltis sp.**	Barbados	–	–	–	–	–	–	–	1	–	–	–	–	1
	NA	–	5	–	–	–	–	–	–	–	–	–	–	5
*Lampropeltis triangulum***	Barbados	–	–	–	–	–	–	–	4	–	–	–	–	4
*Lamprophis fuliginosus***	Barbados	–	–	–	–	–	–	–	2	–	–	–	–	2
*Opheodrys aestivus***	NA	–	–	–	–	12	–	–	–	–	–	–	–	12
*Spalerosophis diadema***	CI	–	–	–	–	–	–	3	–	–	–	–	–	3
*Spilotes pullatus***	Barbados	–	–	–	–	3	–	–	–	–	–	–	–	3
*Thamnophis sirtalis**	NA	–	–	–	–	11	–	–	–	–	–	–	–	11
Naja sp.**	Bahamas	–	–	–	–	–	–	–	–	2	–	–	–	2
*Morelia spilota***	A&B	5	–	–	–	–	–	–	–	–	–	–	–	5
Python molurus (probably *P. bivittatus*)*	Barbados	–	–	–	1	1	–	–	–	–	–	–	–	2
	DR	–	1	–	–	–	–	–	–	–	–	–	–	1
	Jamaica	–	1	–	–	–	–	–	–	–	–	–	–	1
	NA	–	2	–	–	2	–	–	–	4	–	–	–	8
*Python regius**	A&B	500	–	–	–	–	–	–	–	–	–	–	–	500
	NA	–	–	–	–	1	–	–	–	8	4	–	–	13
Chrysemys sp.**	Barbados	–	–	–	–	–	–	25	–	–	–	–	–	25
	CI	–	–	–	–	–	8	–	–	–	–	–	–	8

Species	Country	1998	1999	2000	2001	2002	2003	2004	2005	2006	2007	2008	2009	Total
	NA	–	–	–	–	–	–	–	–	–	1075	–	–	1075
*Cuora amboinensis***	NA	–	–	–	–	20	–	–	–	–	–	–	–	20
*Cyclemys dentata***	Bahamas	–	–	–	–	–	3	–	–	–	–	–	–	3
*Graptemys geographica***	Barbados	–	–	–	–	–	–	–	–	10	–	–	–	10
*Graptemys nigrinoda***	NA	50	–	–	–	–	–	–	–	–	–	–	–	50
*Graptemys pseudogeographica**	NA	–	–	–	–	–	–	–	–	–	925	–	–	925
*Pseudemys floridana***	NA	100	–	–	–	–	100	–	–	–	–	–	–	200
*Pseudemys nelsoni**	NA	200	–	–	–	–	1000	–	–	–	–	–	–	1200
Pseudemys sp.*	Barbados	–	–	–	–	–	–	–	–	505	–	–	–	505
	DR	–	–	–	–	5000	–	–	4000	–	–	–	–	9000
	NA	5000	–	1000	1500	–	500	1000	1000	1200	1100	–	–	12300
	Bahamas	–	–	–	–	12	–	–	–	30	20012	12	–	20066
	Barbados	–	–	–	–	–	100	100	–	70	49	–	–	319
Trachemys scripta	Aruba	–	–	–	50	–	–	6	–	–	–	–	–	56
	DR	1800	4000	11500	15110	10500	9000	6000	8000	10000	16000	6000	–	97910
	Grenada	–	–	–	–	–	1	–	–	–	–	–	–	1
	Jamaica	–	–	100	–	–	–	–	–	–	–	–	–	100
	NA	850	500	–	–	200	2000	–	–	550	1200	4500	–	9800
	PR	–	–	–	–	–	–	–	–	–	–	50	–	50
	T&C	–	–	–	–	–	–	–	–	100	–	–	–	100
Trachemys sp.	Dominica	8	–	–	–	–	–	–	–	–	–	–	–	8
	DR	–	–	–	–	3000	–	–	–	–	–	–	–	3000
	Grenada	–	–	–	–	100	–	–	–	–	–	–	–	100
Apalone ferox	NA	100	–	–	–	–	–	–	–	–	–	–	–	100
Chelonoidis	Barbados	–	–	41	–	–	–	–	–	–	–	–	–	41
(= *Geochelone*) *carbonaria*														
	PR	–	–	–	–	–	–	–	100	300	–	–	–	400
Non-CITES entry (reptiles)	NA	–	–	–	–	4	–	–	–	–	–	–	–	4

Appendix 4. Species imported into the US from the countries indicated (USFWS LEMIS database). Those marked with an asterisk (*) do not occur (native or introduced) in the country of origin. Those marked with a double-asterisk (**) do not occur in the greater Caribbean. BVI = British Virgin Islands, CI = Cayman Islands, DR = Dominican Republic, NA = Netherlands Antilles, T&C = Turks & Caicos Islands.

Species	Country	1998	1999	2000	2001	2002	2003	2004	2005	2006	2007	2008	Total
Rhinella marina	Barbados	–	–	–	–	–	80	–	–	35	–	285	400
(= *Bufo marinus*)	DR	–	200	422	–	–	–	–	–	–	–	–	622
*Hyla cinerea**	Haiti	–	–	–	–	–	–	–	–	–	–	27	27
Hyla sp.	Haiti	–	–	–	222	15	1713	1931	352	316	73	26	4648
Osteopilus vastus (= *Hyla vasta*)	Haiti	–	–	–	–	–	–	–	–	–	489	399	888
Osteopilus sp.	CI	–	–	–	1	–	–	–	–	–	–	–	1
Scinax sp.*	T&C	–	2	–	–	–	–	–	–	–	–	–	2
Eleutherodactylus sp.	Barbados	–	–	–	–	50	–	–	–	484	–	–	534
	Grenada	–	–	–	–	–	–	–	20	–	–	–	20
	Jamaica	30	–	–	–	–	18	–	–	–	–	–	48
	T&C	–	3	–	–	–	–	–	–	–	–	–	3
Leptodactylus fallax	Dominica	–	–	–	7	–	–	–	–	–	–	–	7
*Lithobates catesbeianus**	Dominica	–	6280	–	–	5000	–	–	–	–	–	–	11280
(= *Rana catesbeiana*)	DR	689324	1087716	392363	218788	5218	2797	130000	5500	47930	78296	167550	2825482
	BVI	–	–	–	–	6000	–	–	–	–	–	–	6000
*Rana latastei***	DR	–	–	–	369	–	–	–	–	–	–	–	369
Lithobates sp. (= *Rana* sp.)	DR	35750	9938	781	–	–	–	–	–	–	5550	1590	53609
*Notophthalmus viridescens***	Haiti	–	–	–	–	886	–	–	–	–	–	–	886
Non-CITES entry (amphibian)	DR	–	–	–	1	–	–	–	–	–	–	–	1
	Haiti	–	–	–	335	49	66	160	–	–	–	–	610
	Barbados	–	–	–	–	–	–	–	–	7	–	–	7
	Haiti	–	–	–	–	–	–	6	–	–	–	–	6
Alligator mississippiensis	Bahamas	–	–	–	–	–	–	–	–	–	–	1	1
*Crocodylus siamensis***	Haiti	–	–	–	–	–	–	–	–	–	132	–	132

Species	Country	1998	1999	2000	2001	2002	2003	2004	2005	2006	2007	2008	Total
*Agama agama***	Haiti	–	–	–	–	–	–	–	2537	86	–	–	2623
Uromastyx sp.**	Haiti	400	–	–	–	–	–	–	–	–	–	–	400
Diploglossus sp. (probably *Celestus* sp.)	Haiti	–	–	–	5	20	9	4	–	–	–	–	38
*Chamaeleo bitaeniatus***	BVI	–	–	–	–	–	–	–	–	–	50	–	50
*Chamaeleo dilepis***	BVI	–	–	–	–	–	–	–	–	–	25	–	25
*Chamaeleo ellioti***	BVI	–	–	–	–	–	–	–	–	–	50	–	50
*Chamaeleo hoehnelii***	BVI	–	–	–	–	–	–	–	–	–	75	–	75
Alsophylax sp.**	CI	–	–	–	2	–	–	–	–	–	–	–	2
Gekko sp.*	Barbados	–	–	–	–	–	1	–	–	–	–	–	1
Gonatodes albogularis	Haiti	–	–	–	–	–	–	–	200	–	–	–	200
Gonatodes sp.	Haiti	–	–	–	–	–	–	125	–	125	–	–	250
Hemidactylus mabouia	Barbados	–	–	–	–	12	–	–	–	5	–	–	17
Hemidactylus sp.	Barbados	–	–	–	–	–	–	–	–	12	–	–	12
	CI	–	–	–	4	–	–	–	–	–	–	–	4
	T&C	–	3	–	–	–	–	–	–	–	–	–	3
*Hemidactylus turcicus**	Jamaica	–	–	–	–	–	10	–	–	–	–	–	10
Sphaerodactylus sp.	CI	–	–	–	5	–	–	–	–	–	–	–	5
	T&C	–	9	–	–	–	–	–	–	–	–	–	9
Tarentola sp.*	Haiti	–	–	–	–	–	–	–	50	–	–	–	50
*Anolis carolinensis**	CI	–	–	–	1	–	–	–	–	–	–	–	1
	Haiti	–	–	–	54	–	–	125	–	–	–	–	179
Anolis chlorocyanus	Haiti	–	–	–	–	–	–	–	250	–	–	–	250
*Anolis equestris**	Barbados	–	–	–	–	–	–	–	–	75	–	–	75
Anolis sagrei	CI	–	–	–	5	–	–	–	–	–	–	–	5
	Jamaica	–	–	–	–	–	10	–	–	–	–	–	10
Anolis sp.	Barbados	–	–	–	–	20	22	–	–	100	–	–	142
	CI	–	–	–	5	–	–	–	–	–	–	–	5
	DR	–	–	–	10	–	–	–	–	–	–	–	10
	Haiti	–	–	–	900	–	863	2055	2082	575	245	–	6720
	Jamaica	–	–	121	14	45	–	–	–	–	–	–	180

Species	Country	1998	1999	2000	2001	2002	2003	2004	2005	2006	2007	2008	Total
	T&C	–	3	–	–	–	–	–	–	–	–	–	3
Cyclura nubila	CI	–	4	–	–	–	–	–	10	–	–	–	14
Iguana iguana	CI	–	–	–	–	–	2	–	–	–	–	–	2
	BVI	1	–	–	–	–	–	–	–	–	–	–	1
*Leiocephalus carinatus**	Haiti	–	–	–	3000	3050	13425	15476	4100	2500	7875	7230	56656
Leiocephalus personatus	Haiti	–	–	2260	1136	3270	14100	19422	3850	2650	7461	7500	61649
Leiocephalus schreibersii	Haiti	–	–	4711	3244	–	–	1610	4010	500	–	–	14075
Leiocephalus sp.	Haiti	–	–	6100	6700	–	–	–	–	–	–	–	12800
	T&C	–	3	–	–	–	–	–	–	–	–	–	3
Sceloporus sp.**	Haiti	–	–	–	–	–	53	25	–	–	–	–	78
Lacerta sp.**	NA	–	9	–	–	–	–	–	–	–	–	–	9
Mabuya sp.	T&C	–	3	–	–	–	–	–	–	–	–	–	3
Ameiva ameiva	Barbados	–	–	–	–	–	–	–	–	22	–	1	23
Ameiva sp.	Jamaica	3	–	–	–	–	–	–	–	–	–	–	3
Boa constrictor	Aruba	–	–	–	–	–	2	–	–	–	–	–	2
Epicrates chrysogaster	T&C	–	1	–	–	–	–	–	–	–	–	–	1
Epicrates fordii	Haiti	–	–	–	40	–	28	–	–	–	–	–	68
Epicrates gracilis	Haiti	–	–	–	91	–	9	–	–	–	–	–	100
Epicrates striatus	Haiti	–	–	–	66	–	23	226	88	–	–	–	403
Lampropeltis sp.**	DR	–	1	–	–	–	–	–	–	–	–	–	1
Naja sp.**	Bahamas	–	–	–	–	–	–	–	–	2	–	–	2
*Python molurus** (probably *P. bivittatus*)	Aruba	–	–	–	–	–	–	–	–	–	1	–	1
Tropidophis greenwayi	T&C	–	1	–	–	–	–	–	–	–	–	–	1
Tropidophis haitianus	Haiti	–	–	–	15	–	42	396	90	–	–	–	543
Typhlops sp.	T&C	–	2	–	–	–	–	–	–	–	–	–	2
Bothrops sp.	St. Lucia	–	–	–	2	–	–	–	–	–	–	–	2
Chrysemys sp.*	Cuba	–	–	–	1	–	–	–	–	–	–	–	1
*Mauremys caspica***	Cuba	–	–	1	–	–	–	–	–	–	–	–	1
Pseudemys sp.*	Cuba	1	–	–	2	–	–	–	–	–	–	–	3
Terrapene sp.*	NA	–	2	–	–	–	–	–	–	–	–	–	2

Species	Country	1998	1999	2000	2001	2002	2003	2004	2005	2006	2007	2008	Total
Trachemys scripta	CI	–	–	–	–	–	4	–	–	–	–	–	4
	DR	3000	–	–	–	–	2	–	–	–	–	–	3002
	Jamaica	–	–	–	–	–	–	–	–	–	2	–	2
Trachemys sp.	Cuba	–	–	–	1	–	–	–	1	–	–	–	2
Chelonoidis carbonaria	Barbados	98	–	20	451	20	40	–	–	8	1	25	663
(= *Geochelone carbonaria*)	BVI	–	–	–	–	–	–	–	–	–	400	–	400
*Pyxis arachnoides***	Martinique	–	–	–	–	–	–	2	–	–	–	–	2
Non-CITES entry (reptile)	Barbados	–	–	–	–	–	–	–	–	9	–	–	9
	DR	–	–	–	–	2724	–	–	–	–	–	–	2724
	Haiti	–	–	–	20	3	10	5	3	–	–	–	41
	T&C	–	1	–	–	–	5	–	–	–	–	–	6

Appendix 5. Species exported from the shown country, imported to the US, then re-exported mostly to European, Canadian, some Asian markets, and a few are re-exported to countries in the greater Caribbean (USFWS LEMIS database). Those marked with an asterisk (*) do not occur (native or introduced) in the country of origin. Those marked with a double-asterisk (**) do not occur in the greater Caribbean. DR = Dominican Republic, NA = Netherlands Antilles, PR = Puerto Rico, T&C = Turks & Caicos Islands, USVI = US Virgin Islands.

Species Exported to the US	Country	1999	2000	2001	2002	2003	2004	2005	2006	2007	2008	Total
Peltophryne (= *Bufo*) *lemur*	PR	1	–	–	–	–	–	–	–	–	–	1
Rhinella marina (= *Bufo marinus*)	Barbados	–	–	–	–	–	–	–	–	–	6	6
	DR	146	80	–	–	–	–	–	–	–	–	226
Hylidae (listed as *Hyla* sp.)	Bahamas	–	–	32	–	–	–	–	–	–	–	32
	Haiti	–	–	–	3	85	151	27	23	3	6	298
Osteopilus vastus (= *Hyla vasta*)	Haiti	–	–	–	–	–	–	–	–	56	30	86
Osteocephalus sp.**	Cuba	–	–	–	–	–	–	–	–	–	6	6
Phyllomedusa sp.**	PR	2	–	–	–	–	–	–	–	–	–	2
*Tlalocohyla loquax***	Haiti	–	–	–	–	–	–	–	–	4	–	4
*Cynops orientalis***	Haiti	–	–	–	–	–	–	100	–	–	–	100
Reptile (?)	Haiti	–	–	–	–	–	3	–	–	–	–	3
*Agama agama***	Haiti	–	–	–	–	–	40	2950	300	–	–	3290

Species Exported to the US	Country	1999	2000	2001	2002	2003	2004	2005	2006	2007	2008	Total
Japalura sp.**	Haiti	–	–	–	–	–	6	–	–	–	–	6
*Physignathus cocincinus***	USVI	–	–	–	50	–	–	50	–	–	–	100
Diploglossus sp. (probably *Celestus* sp.)	Haiti	–	–	–	–	2	3	–	–	–	–	5
*Eublepharis macularius***	Haiti	–	–	–	–	500	–	–	–	–	–	500
*Gekko gecko**	USVI	–	–	12	–	–	–	–	–	–	–	12
Gonatodes albogularis	Haiti	–	–	–	–	–	–	69	–	–	–	69
Gonatodes sp.	Haiti	–	–	–	–	–	65	30	–	–	–	95
Hemidactylus sp.	Haiti	–	–	–	–	–	–	100	–	–	–	100
*Ptychozoon kuhli***	Haiti	–	–	–	–	–	–	24	–	–	–	24
Tarentola sp.*	Haiti	–	–	–	–	–	–	52	–	–	–	52
*Anolis bimaculatus**	Haiti	–	–	–	–	–	–	30	–	–	–	30
*Anolis carolinensis**	Haiti	–	–	–	–	–	3	25	–	–	–	28
Anolis chlorocyanus	Haiti	–	–	–	–	–	97	169	31	58	10	365
*Anolis roquet**	Haiti	–	–	–	–	–	–	–	–	–	38	38
Anolis sp.	Haiti	–	–	–	–	97	835	273	32	70	91	1398
	Jamaica	–	–	3	–	–	–	–	–	–	–	3
Leiocephalus carinatus	Bahamas	–	–	–	–	11	57	–	–	–	–	68
*Leiocephalus carinatus**	Haiti	–	125	937	934	3215	4497	1030	25	637	318	11718
*Leiocephalus personatus**	Bahamas	–	–	5	–	–	–	–	–	–	–	5
Leiocephalus personatus	Haiti	6	150	163	12	1662	3728	989	1227	3546	4068	15551
*Leiocephalus schreibersii**	Bahamas	–	–	–	–	200	–	–	12	–	–	212
Leiocephalus schreibersii	Haiti	–	237	296	126	2051	4156	1461	1215	2367	2776	14685
Leiocephalus sp.	Haiti	–	20	765	59	1024	902	598	414	86	136	4004
Polychrus sp.**	PR	4	–	–	–	–	–	–	–	–	–	4
*Takydromus sexlineatus***	Haiti	–	–	–	–	–	–	100	–	–	–	100
*Riopa fernandi***	Haiti	–	–	–	–	–	–	–	–	–	25	25
Epicrates striatus	Haiti	–	–	–	–	–	42	2	–	–	–	44
*Lampropeltis getula***	Haiti	–	–	–	–	–	–	–	–	–	1	1
*Opheodrys aestivus***	Haiti	–	–	–	–	–	69	–	–	–	–	69
*Pseustes poecilonotus***	NA	–	–	–	–	–	2	–	–	–	–	2

Species Exported to the US	Country	1999	2000	2001	2002	2003	2004	2005	2006	2007	2008	Total
Thamnophis sauritus	Bahamas	–	–	1	–	–	–	–	–	–	–	1
Tropidophis haetianus	Haiti	–	–	–	–	–	34	5	–	–	–	39
*Terrapene carolina**	Montserrat	–	–	–	–	–	–	–	–	3	–	3
*Terrapene ornata***	Montserrat	–	–	–	–	–	–	–	–	2	–	2
Chelonoidis (= *Geochelone*) *carbonaria*	Barbados	–	–	10	–	–	–	2	–	–	–	12
*Testudo horsfieldii***	Montserrat	–	–	–	–	–	–	–	–	5	–	5
Non-CITES entry (reptile?)	Haiti	–	–	6	–	–	54	1	37	–	–	98
Rhinella marina (= *Bufo marinus*)	PR	–	–	–	1	–	–	–	–	–	–	1
Afrixalus sp.**	PR	12	–	–	–	–	–	–	–	–	–	12
Phyllomedusa sp.**	PR	9	–	–	–	–	–	–	–	–	–	9
Eleutherodactylus sp.	PR	–	50	–	–	–	–	–	–	–	–	50
Lithobates catesbeianus (= *Rana catesbeiana*)	DR	1450	–	–	–	–	–	–	–	–	–	1450
Lithobates sp. (= *Rana* sp.)	DR	–	–	–	–	–	–	–	–	1260	–	1260
*Cynops orientalis***	St. Lucia	–	–	100	–	–	–	–	–	–	–	100
Diploglossus sp. (probably *Celestus* sp.)	Haiti	–	–	–	–	–	–	1	–	–	–	1
Anolis sp.	Jamaica	–	4	–	–	–	–	–	–	–	–	4
*Leiocephalus carinatus***	Haiti	–	–	25	–	–	–	–	–	–	–	25
Leiocephalus schreibersii	Haiti	–	–	50	–	–	–	–	–	–	–	50
Leiocephalus sp.	Haiti	–	–	25	–	–	–	–	–	–	–	25
Atheris sp.**	T&C	–	–	–	–	11	–	–	–	–	–	11
Trachemys decussata	Jamaica	–	–	–	–	–	1	–	–	–	–	1
Trachemys stejnegeri	DR	–	–	–	–	–	1	–	–	–	–	1
Chelonoidis (= *Geochelone*) *carbonaria*	Barbados	–	–	–	–	–	–	1	–	–	–	1
Non-CITES entry (reptile?)	NA	–	–	–	–	13	–	–	–	–	–	13

Conservation of amphibians and reptiles in Aruba, Curaçao and Bonaire

Gerard van Buurt

Kaya Oy Sprock 18, Curaçao
e-mail: gvanbuurt@gmail.com

Abstract. Curaçao and Bonaire form part of the Netherlands Antilles, while Aruba has a "status aparte" within the Kingdom of the Netherlands. Áll three islands are relatively arid compared to a typical Caribbean island, with mean annual rainfall of 409-553 mm, and experience several periods of drought lasting two or more years each century. A short history of the islands is given, and protected areas are described. The laws and regulations protecting amphibians and reptiles are complex, with general laws originating from the Kingdom of the Netherlands participation in international conventions (such as CITES) together with supplemental laws of the Netherlands Antilles and individual islands. Sea turtles are generally well protected, although their nesting beaches would be vulnerable to a rise in sea level. Among the terrestrial herpetofauna, only the Aruba Island rattlesnake (*Crotalus unicolor*) is on the IUCN Red List, being Critically Endangered. The status of this species and others of particular interest is described. The Curaçao Island snake (*Liophis triscalis*) should probably be included as Vulnerable or even Endangered, though there is insufficient information at present. *Iguana iguana* populations on the different islands, and the Curaçao whiptail (*Cnemidophorus murinus murinus*) on Klein Curaçao, are distinctive and significant for conservation. An overview is given of introduced amphibians and reptiles and their possible effects on the native fauna. The arid climate of the islands may hinder the establishment of invasive species, which are often not able to survive in the bush and thus reduces their impact on native species.

Key words: *Boa constrictor*; *Cnemidophorus murinus murinus*; *Crotalus unicolor*; drought; invasive species; Klein Curaçao; *Liophis triscalis*; Netherlands Antilles; *Osteopilus septentrionalis*; sea turtles.

Introduction

General ecology

The islands of Aruba, Curaçao and Bonaire lie in the southern Caribbean Sea. Their climate is rather arid compared to most Caribbean islands, with mean annual rainfall of 409, 553, and 463 mm respectively (Meteorological Service of Netherlands Antilles and Aruba, average 1971-2000), and mean temperature of about 28°C. The hilly areas receive slightly more rain, and rainfall can be quite variable from year to year. Dry years have only 200-300 mm of rain, while the maximum is about 1100 mm. The dry season normally runs from March to June, and October and November have most rain; a dry year has a longer dry season. The period 1830-2004 experienced 12 extended dry periods; eight of 2 years, three of 3 years, and one of 4 years (1902-1905, when food importation from Venezuela was necessary). The last droughts were in 1986-1987 and 2001-2002. Mean annual rainfall was only 341 mm in Curaçao during these 12 extended dry periods.

The vegetation is typically xeric with candelabra cacti, opuntias, and acacias. At the end of long dry periods even the opuntia cactus starts to shrivel, the bush loses its leaves and becomes "transparent", and emaciated iguanas cling to trees or fall on the ground. Soils are either of volcanic origin or calcareous — the latter can store water better and have a somewhat different vegetation. Aruba has lower rainfall and is more arid than the other islands, but also has less permeable soil so that pools of surface water (tankis) last much longer there.

Aruba, Curaçao and Bonaire together with the Venezuelan islands of Los Monjes, Islas Aves, Los Roques, La Orchila and La Blanquilla form an island archipelago north of the Venezuelan coast. Biogeographically, these islands do not belong to the West Indian region; their flora and fauna are mainly of South American origin (Wagenaar Hummelinck, 1940). However, there are many endemic and some West Indian elements, which justify the view that these islands form a small but distinct zoogeographical sub-region belonging to the South American realm. Several organisms introduced in the past turned out to be invasive species, some of which had detrimental effects on the local flora and fauna (van Buurt, 1999).

History and political structure

For general information on the history of Aruba, Curaçao and Bonaire see Hartog (1957, 1961, 1968). The earliest human remains on the islands, dating from 4500 years Before Present, were found in Curaçao. On the other islands the earliest are 4000 BP in Aruba and 3300 BP in Bonaire. These people are now termed Archaic Indians (Haviser, 1987). The Caquetío Indians, a tribe belonging to the Arawak language family, reached the islands from South America around 500 AD. The group "indios Aruba" were living in Aruba and the "indios Curaçao" in Curaçao and Bonaire. In 1499 the Spanish came to the islands, and the population was raided to get slaves for mines on Hispaniola. In 1634 the islands were captured by the Dutch who used Curaçao as a naval base to raid Spanish shipping and obtained salt, dyewood (*Heamatoxylon brasiletto*) and some "lignum vitae" (*Guaiacum officinale*). Later when the Dutch obtained the Asiento (a contract to deliver African slaves to the Spanish main), the island became a slave depot. Aruba was used as a horse farm by the Dutch West-India Company, and Bonaire produced salt. The introduction of slaves from Africa led to the development of Papiamentu, a Creole language with an African grammar with Spanish, Portuguese, Dutch and in more recent times also English words. The islands were occupied by the British from

1806 to 1816. During the colonial period the slave trade, smuggling, and salt were the most important economic activities, and some agriculture existed. Sugar was planted for local use in Curaçao, and to produce rum, but being too arid the island never exported sugar. Thus, it was not necessary to introduce the cane toad or the mongoose. In the 17th century indigo (*Indigofera tinctoria*), and in the late 19th and early 20th century aloë (*Aloe vera*) and sisal (*Agave sisalana*), were cultivated for export, but the islands were very poor (Renkema, 1981). In 1916 the Shell oil company started a refinery on Curaçao, and Standard oil of New Jersey started a refinery a few years later in Aruba, both to process Venezuelan oil, thus starting the modern age. In 1954 the islands became autonomous, and in 1986 Aruba seceded from the Netherlands Antilles; the "status aparte". Nowadays tourism, the service sector, offshore banking, transshipment and bunkering, and oil refineries form the mainstay of the economy. Bonaire is almost totally dependant on tourism, especially dive-tourism.

Currently the Netherlands Antilles consist of the islands of Curaçao and Bonaire, together with the Windward Islands of St. Maarten (Dutch part), Saba and St. Eustatius. Aruba is a separate autonomous entity within the Kingdom of the Netherlands. Political negotiations on the future status of the other islands are presently being held. It has been agreed that these should result in a "status aparte" for Curaçao and St. Maarten, while the other islands Saba, St. Eustatius and Bonaire will maintain a more direct link with the Netherlands.

Nature Conservation Legislation

Legal system, constitution and nature conservation

According to the "Eilandenregeling Nederlandse Antillen", 1954, which forms part of the present constitution of the Netherlands Antilles, nature conservation is a responsibility of the island governments. However, international treaties are a central government function, and in practice there is often some overlap of responsibilities. The same holds true for fisheries. Consequently the central government of the Netherlands Antilles has enacted general legislation to comply with international treaties, but where not constrained by treaties the islands are free to enact (or not to enact) their own nature regulations. Thus in addition to the central government laws, most islands have their own nature and/or fisheries regulations, which were often copied from one to another and thus similar, but in some cases quite different. Aruba as a separate autonomous entity is not hindered by these constitutional constraints, and has combined all nature conservation into one piece of legislation.

Treaties which are of significance for the conservation of amphibians and reptiles, of which the Netherlands Antilles (NA) and/or Aruba (A) are a party through the Kingdom of the Netherlands (N) are:

- CITES (N, NA, A).
- Cartagena Convention (NA, A) and SPAW protocol (NA).

- Inter-American Convention for the Protection and Conservation of Sea Turtles (NA).
- Convention on Biological Diversity (N, NA, A).
- Convention on Migratory species or Bonn Convention (N, NA, A).
- RAMSAR Convention (N, NA, A).

National legislation of a general nature

In the Netherlands Antilles the most important piece of nature legislation is the National Nature Conservation Ordinance "Landsverordening grondslagen natuurbeheer en bescherming" (PB 1998, 49; modified PB 2001, 41). The full text of this and other legislation described below (in Dutch, some with English translation) is available at the website: http://www.mina.vomil.an. This law follows CITES and the SPAW protocol and is the national legislation which enacts these treaties. In some respects this law goes further than CITES, which is applicable only to trade across national borders. Species listed under CITES Appendix I and the Annexes I and II of the SPAW protocol, and other species listed in the other treaties mentioned above, are automatically protected and cannot be killed or kept without a permit. Since the law refers to the CITES and SPAW protocol lists, new species added to these lists are automatically protected.

There is a national register of all captive specimens listed under CITES Appendix I, including captive native species. Illegal animals are confiscated and their holder is fined and/or jailed. Legal animals were either registered when the law was enacted, the offspring of registered animals, or were legally imported with proper CITES documentation. Aruba has the almost identical "Natuurbeschermingsverordening" (AB 1995, 2; modified AB 1997, 34) which includes species listed under CITES Appendix I and II. Thus in Aruba all sea turtles, *Iguana iguana*, and *Crotalus unicolor* are protected.

Fisheries laws

Sea turtles in federal waters outside the 12 mile zone are protected under the Netherlands Antilles Fisheries Ordinance or "Visserijlandsverordening" (PB 1991, 74) and the National Fisheries Decree or "Visserijlandsbesluit" (PB 1992, 108), which contains the actual fisheries regulations. These regulate fisheries in the waters of the Exclusive Fisheries Zone from ships larger than 12 m or 6 gross registered tons. Aruba has the almost identical "Visserijverordening (Aruba)" (AB 1992, 116; modified AB 1997, 34).

Island regulations on Curaçao and Bonaire

Other amphibians and reptiles not listed under CITES Appendix I are only protected under separate island laws. Thus on Curaçao and Bonaire *Iguana iguana* is not protected. Sea turtles within the 12 mile zone are protected under island regulations on all of the Netherlands Antilles. In Curaçao sea turtles and their nests are protected under the "Eilandsbesluit bescherming zeeschildpadden" (PB 1996, 8) which is part of the Reef Management Ordinance "Rifbeheerverordening" (PB 1976, 48). In Bonaire sea turtles within the 12 mile zone are protected under the "Eilandsverordening Marien Milieu" ($A^0$1991, 8; modified $A^0$2001, 13).

Protected areas

Parke Nacional Arikok (Aruba). This park has a surface area of 34 km^2 and was established in 1997. A large part of the range of *Crotalus unicolor* is contained within its boundaries (Reinert et al., 2002). Persons who find rattlesnakes near their homes can call the park snake catchers who will release the snakes in the park. There are also some sea turtle nesting beaches within the park.

Christoffelpark and Shete Boca Park (Curaçao). The former is a large terrestrial park in the northwestern part of Curaçao, established in 1978. The adjoining Shete Boca Park runs along the northwestern coast, and has some loggerhead and hawksbill turtle nesting beaches. Most of the lands belong to the Curaçao island government and are managed by an NGO.

Washington and Slagbaai Park (Bonaire). A large park covering the northwestern part of Bonaire, established in 1968 and expanded in 1979. This park includes several sea turtle nesting beaches. Although this park was not instituted by island legislation it is nevertheless effectively protected by conditions that are incorporated into the title deeds to the lands, established by the will of one of the former owners and by the funding agencies which financed the purchase of additional lands.

Klein Bonaire (Bonaire). This 7 km^2 uninhabited island was added to the Bonaire Marine Park in 2001. There are nesting beaches of hawksbill and loggerhead turtles on the north and west coasts of the island. The island is protected under the "Eilandsverordening Marien Milieu" and also by conditions in the title deeds, established by the agencies which financed the purchase of the island.

Sea Turtles

Hawksbills, loggerheads, green turtles and leatherbacks are found around the islands. The olive ridley (*Lepidochelys olivacea*) is an occasional visitor or vagrant (Sybesma and Hoetjes, 1992). Hawksbills, loggerheads, green turtles nest on all the islands, including Klein Curaçao and Klein Bonaire. The leatherback (*Dermochelys coriacea*) nests on Aruba and has occasionally nested in Curaçao (Knip) and Bonaire (Lagun bay). In Aruba the nests of leatherbacks and other turtles are monitored and protected. These activities are part of the activities described in the WIDECAST/Sea Turtle Recovery Action Plan for Aruba (Barmes et al., 1993).

A similar WIDECAST/Sea Turtle Recovery Action Plan exists for the Netherlands Antilles (Sybesma, 1992).

An overview of nesting records for the Netherlands Antilles is given in Debrot et al. (2005). In recent years, turtles are seen more frequently than in the past (pers. obs.). Up to the early 1960's they were fairly common; in the 1970's they became increasingly rare, but recently populations seem to be recovering. This suggests that their legal protection (Bonaire in 1991, Netherlands Antilles in 1992, Aruba in 1995, Curaçao in 1996) and the regional projects to conserve them have had results. In Bonaire the organisation Sea Turtle Conservation Bonaire (formerly Sea Turtle Club Bonaire) has been instrumental in promoting sea turtle conservation through educational campaigns, cleaning debris from beaches and a satellite-tracking program in cooperation with WIDECAST (see www.bonaireturtles.org).

Rising sea levels will affect turtle nesting beaches. A study by Fish et al. (2005) on Bonaire indicates that a 0.5 m rise in sea level could lead to a 32% loss of existing sea-turtle nesting habitat. Some beaches will simply shift landward and new nesting habitat will be formed. This is not possible where the landward movement of the beach is impeded by physical barriers, or will result in changing characteristics of the beach, and thus a net loss of nesting habitat.

Terrestrial Reptiles of Special Conservation Interest

Table 1 lists the terrestrial amphibians and reptiles of Aruba, Curaçao and Bonaire, and the lesser islands of Klein Curaçao and Klein Bonaire, with their status as endemic, native or introduced. Those species of special conservation interest (all reptiles) are considered in detail below.

Aruba Island rattlesnake (Crotalus unicolor)

This is the only species of terrestrial herpetofauna on the islands currently included on the IUCN Red List (IUCN, 2004), where it is classed as Critically Endangered. In 1992 the American Zoo and Aquarium Association in cooperation with the Department of Agriculture and Fisheries in Aruba (DLVVM) published a Conservation Action Plan for the Aruba Island rattlesnake, updated in 1995 (Reinert et al., 1995). At the time field studies suggested a population of less than 225 adults, and indicated that mortality was quite high. Snakes were killed by goats, donkeys, cats, humans and cars. One of the results of the Conservation Action Plan was the establishment of the Parke Nacional Arikok in 1997. In recent years it appears that the population has recovered somewhat, probably as a result of the protection afforded by the new park, increased public awareness, and other conservation measures (Dr. H.K. Reinert, pers. comm.). DLVVM, the Aruba Veterinary Service and Parke Nacional Arikok also support research and projects to educate the general public.

Table 1. Terrestrial amphibians and reptiles of Aruba, Curaçao and Bonaire. E = endemic; Ep = E pluralis, endemic to more than one island; I = Introduced; N = Native; pN = probably native.

	Aruba 190 km^2	Curaçao 444 km^2	Klein Curaçao 1.2 km^2	Klein Bonaire 7 km^2	Bonaire 282 km^2
Amphibians					
Pleurodema brachyops	N	I		I	I
Osteopilus septentrionalis					I
Eleutherodactylus johnstonei	I	I			I
Bufo marinus	I				
Reptiles					
Gonatodes antillensis	I	N	N	N	N
G. albogularis albogularis		I			
G. vittatus vittatus	N				
Phyllodactylus martini		Ep		Ep	Ep
Phyllodactylus julieni	E				
Hemidactylus mabouia	I	I		I	I
Thecadactylus rapicauda	N	N			N
Iguana iguana	N[1]	N[1]		N[1]	N[1]
Anolis lineatus	Ep	Ep	I[2]		
Anolis bonairensis				Ep	Ep
Ameiva bifrontata	pN				
Cnemidophorus arubensis	E				
Cnemidophorus lemniscatus lemniscatus	I				
Cnemidophorus murinus murinus		Ep	Ep[3]		
Cnemidophorus murinus ruthveni				Ep	Ep
Gymnophthalmus lineatus			N	?	N
Gymnophthalmus speciosus	N				
Tretioscincus bifasciatus	N				
Liotyphlops albirostris		N			
Leptotyphlops albifrons					I
Ramphotyphlops braminus	I				
Boa constrictor	I				
Leptodeira bakeri	N				
Liophis triscalis		E			
Crotalus unicolor[4]	E				

[1] Distinct populations.
[2] Introduced 2 November 2005 by Dr. A.O. Debrot, CARMABI.
[3] Klein Curaçao population distinct.
[4] Critically Endangered.

Curaçao Island snake (Liophis triscalis)

The Curaçao Island snake must certainly be considered a highly vulnerable species, possibly even endangered, although it is not included on the IUCN Red List as no population studies exist on which to base such judgment. Many people still kill it, even though many know it is not dangerous. Politically it will be difficult to give this snake legal protection; an educational campaign would certainly be needed to create support first. If the island develops further and more habitat is lost, its survival could well depend on the national park and other protected areas. This snake could also be vulnerable to an invasive species such as the corn snake (below) or African dwarf hedgehog (recently sold in local pet shops).

Cnemidophorus murinus murinus *on Klein Curaçao*

The 1.2 km^2 coral island of Klein Curaçao lies about 9 km southeast of Curaçao. A population of *C. m. murinus* is found there that differs from that on the main island (pers. obs.). Both sexes are smaller, and there is greater sexual size dimorphism. The colors of the males are more intensely blue-grey and the line near the eye in the males is usually broken up in a pattern of dots (fig. 1). The population is in good shape. The island receives a maximum of about 600 visitors a week and the lizards congregate near people and receive supplementary food. Goats were removed from the island in January 1998 and the vegetation, although sparse, has recovered considerably. Since *C. murinus* is a largely herbivorous lizard, the removal of goats has eliminated much competition for food. A population of feral cats was extirpated by Dr. A.O. Debrot.

Even so, any population of lizards found only in such a small area must be considered vulnerable. Most of the island is less than 2 m above sea level, with a maximum of 4.7 m in the east where there is a ridge of coral rubble. In the past the island had a low limestone ridge with a maximum height of about 7 m, but the island was mined for phosphates from 1871 to 1913 and the ridge was levelled (Stienstra, 1991). Klein Curaçao lies at the southern edge of the hurricane range and it is known that the island was completely swept by waves during the large hurricane of 1877 (Hendrikse, 2005). Part of the higher ridge still existed at the time. With sea levels rising due to global warming, a similar hurricane could be even more damaging now. Nevertheless, many even smaller keys with populations of *Cnemidophorus nigricolor* exist in Las Aves and Los Roques (Venezuela), that must have been subject to similar hurricanes in the past, yet these populations apparently survived.

Iguana iguana *on Aruba, Curaçao and Bonaire*

The iguanas on these islands are considerably smaller than those from the mainland, not exceeding 1 m in length (Marken Lichtenbelt and Albers, 1993; Marken Lichtenbelt et al., 1993), while the mainland iguana can reach a length of about 1.7 m. They are much less arboreal and behave more like ground iguanas. There are some differences between the iguanas on Aruba, Curaçao and Bonaire. In Aruba many individuals are found with the striped pattern on both the body and tail well developed; in Bonaire the stripes on the body are subdued and not clearly visible; while those from Curaçao tend to be in between (van Buurt, 2001, 2005). In the Curaçao iguanas dominant males have substantial blue on the head.

Figure 1. Upper: young male *Cnemidophorus murinus murinus* from Curaçao — note the lines on the head. Lower: male *C. m. murinus* from Klein Curaçao — note the dots on the head. These males also have a more bluish-grey colour (colour originals — see www.ahailey.f9.co.uk/appliedherpetology/cariherp.htm).

Iguanas are protected in Aruba, and are plentiful, almost garden pests. Iguanas are not protected on Curaçao and Bonaire, but are quite common and certainly not endangered. It is forbidden to use catapults, air guns and firearms on government lands, and firearms are strictly regulated. Nowadays iguanas are hunted much less than in the past when the islands were very poor and iguanas were needed as food.

Introduced Amphibians and Reptiles

Species of amphibians and reptiles introduced on one or more of the islands are included in table 1. Their effects on the indigenous species, as far as can be ascertained, are described below.

Amphibians

Whistling frog. *Eleutherodactylus johnstonei* was imported with plants from Venezuela (van Buurt, 2001, 2005), to Curaçao in the 1970s, later on Aruba and Bonaire. It is dependant on gardens; the bush (locally called mondi) is too arid. Since the whistling frog cannot penetrate into the mondi, local species are not threatened by its presence.

Cuban treefrog. *Osteopilus septentrionalis* (fig. 2) has recently been introduced in Bonaire (van Buurt, 2005), with plants from Florida. It is too early to tell whether these frogs are also dependant on gardens. Since they climb trees they may be a threat to the endemic *Anolis bonairensis* and may also compete with them for food. They almost disappeared in the 2005 dry season, but some surely found refuge in gardens. There are few cisterns left since the advent of piped water, and there is now almost no standing fresh water to be found in the dry season. It is known that such cisterns are used by the Cuban tree frog in other islands (Meshaka, 2001). Bonairians were indignantly complaining that the Cuban tree frog had been seen to eat their local "Sapu di Boneiru" (Bonaire frog) *Pleurodema brachyops*. This frog was introduced in Bonaire in 1928 (below); however most people consider it a local species. *Pleurodema* is adapted to an arid environment and can survive two or three years of drought. It remains to be seen whether the Cuban tree frog can be a threat to it in drier areas away from gardens.

Cane toad. *Bufo marinus* was introduced to Aruba in the 1960's from Colombia (van Buurt, 2001, 2005). It has negatively affected the populations of the local *Pleurodema brachyops*. The fact that it managed to establish itself on an arid island like Aruba indicates that it could also establish itself on Curaçao and/or Bonaire.

Four-eyed frog. *Pleurodema brachyops* was introduced in Curaçao and Bonaire from Aruba where it is native (Wagenaar Hummelinck, 1940). It seems to have occupied an empty ecological niche in these islands, and does not threaten any local species. In Curaçao it is now an extra source of food for *Liophis triscalis*. With the introduction of *Bufo marinus* on Aruba, *Pleurodema brachyops* is now more common in Curaçao and Bonaire than in its native Aruba (van Buurt, 2005).

Figure 2. Cuban tree frog *Osteopilus septentrionalis* on Bonaire. The Netherlands Antilles guilder has a diameter of 24 mm.

Reptiles

Cosmopolitan house gecko. *Hemidactylus mabouia* was introduced in Curaçao in the late 1980's; it is now found in houses all over the island. In Bonaire it came in around 2000, and in Aruba it was first reported (by Mikael Lundberg) in January 2002 (van Buurt, 2005). It is found exclusively in houses and gardens where it has displaced *Phyllodactylus martini* (in Curaçao and Bonaire) and to a certain extent also *Gonatodes antillensis*. It seems that *H. mabouia* cannot penetrate into the mondi or displace *P. martini* in this environment, but the reason is not apparent.

Rainbow whiptail. *Cnemidophorus lemniscatus lemniscatus* was first found in Aruba in 1953 (Lammarée, 1970). It is found only in the south-eastern part of the island, where it is not common. It competes with the Aruba Island Whiptail (*Cnemidophorus arubensis*), which also occurs throughout the same area and is very common. Within its general range the rainbow whiptail is found only in areas with a more sandy soil, and seems able to compete only in these areas (van Buurt, 2005).

Boa (*Boa constrictor*). Boas have been regularly caught on Aruba since 1999, and in subsequent years their numbers have increased considerably. 273 *B. constrictor* were reported to and captured by personnel of the Arikok National Park and the DLVVM between April 1999 and December 2003. The Aruba Veterinary Service is also involved in efforts to eradicate boas and protect the Aruba Island rattlesnake. *Boa constrictor* is now an established species on Aruba (Quick et al., 2005). It is likely that they arrived with the pet trade, as was the case in Curaçao, where in previous years boas were reported quite regularly, but fortunately did not manage to establish themselves (van Buurt, 2005). In recent years CITES regulations started to be enforced by customs and nowadays sightings of boas are rare in Curaçao. In Aruba some specimens with white blotches are found, which points to a pet ancestry. The boa is very likely a competitor to the Aruba Island rattlesnake, and a new predator for the iguana. Planned research by Dr. Howard Reinert (College of New Jersey, USA) and R. Andrew Odum (Toledo Zoological Society) will follow boas by telemetry and use the information obtained to trap them more effectively.

Corn snake. *Elaphe guttata* nearly became established in Curaçao (Perry et al., 2003). Since 2001 corn snakes from North America were found on both Curaçao (2001, 2002 and 2003) and Bonaire (2002). In Curaçao they were all in an area called Kwartje where plants imported from Florida were stored. The snakes could also have been escaped or released pets as corn snakes had been sold by a local pet shop. Some were trapped and killed in February 2004 but no more sightings have been reported from the area since. Even so, we cannot be sure they were extirpated as corn snakes were almost certainly reproducing, several juveniles being found. It is conceivable that the Corn snake could be a serious threat to the endemic Curaçao snake. It is strange that a species with a normal range in temperate zones can survive and reproduce in a harsh purely tropical environment, although corn snakes are established on Anguilla which is fully tropical. They have also been reported from St. Thomas (US Virgin islands), Antigua and St. Barth's (Perry et al., 2003). Compared to these islands Curaçao is a more arid and a warmer island. Possibly the snakes came from southern Florida and were already adapted to a more or less tropical climate.

Flowerpot blindsnake. Several individuals of *Rhamphotyphlops braminus* were found at and near a plant importer in Aruba, reported by Mr. Facundo Franken of the DLVVM. The first snake was found in 2003 in an area called Bakval, near Malmok. Later on several more were encountered. This introduction probably will not have a detrimental effect as Aruba does not have any native typhlopid with which it could be in competition.

Discussion

Protecting against invasive species

The Department of Agriculture and Fisheries in Curaçao has developed a PowerPoint presentation giving an overview of past introductions and their effects. More importantly, it describes the "alert species"; those that can pose a new threat to the flora and fauna of the islands, and which could conceivably come in. This presentation is used to increase awareness of this problem and is given to Customs, Aerocargo and Department of Agriculture personnel, importers of plants, nature groups and the public in general. Examples of such "alert species" that could become established and which are likely to be a problem are; the opuntia cactus borer (*Cactoblastis cactorum*), agave borer weevils (*Scyphophorus acupunctatus*), fire ants (*Solenopsis invicta*), *Ganoderma zonatum* fungus on palms, and others. An "alert list" also includes species which have already established themselves on one of the islands, but can still be kept out of the others.

The "alert list" includes the following amphibians and reptiles: the cane toad, the Cuban tree frog, the boa, and the corn snake. Usually species are most affected by new species that compete with them, or by new predators. In the case of amphibians and reptiles, the most likely threats would often be other amphibians and reptiles. Of course, species like fire ants, if introduced, could also pose a threat to local herpetofauna. Then there is always the unknown; any new introduction could potentially have totally unforeseen consequences which could also affect amphibian and reptile populations. In my opinion new invasive species are potentially the most serious threat to local indigenous species of amphibians and reptiles. However, regulations regarding exotic species and/or "alert species" are still lacking on all three islands.

Influence of prolonged dry periods

Even though some parts of the vegetation of Aruba, Curaçao and Bonaire look very similar to the drier parts of some of the West Indian islands, especially during the rainy season, they are in fact much more arid. Not only is rainfall lower, the rate of evaporation is higher and extended dry spells occur. For example in St. Maarten the average annual rainfall is 1047 mm. The driest year on record (since 1879) is 1967, with 689 mm, which is still above the 553 mm average and more than three times the driest year in Curaçao (with 208 mm). In Anguilla, an island where both the Cuban treefrog and the corn snake became established (Hodge et al., 2003) the average rainfall is 1089 mm. The Cuban treefrog is also established on St. Maarten, (Powell et al., 2005). In these islands there is nothing comparable to the long dry periods occurring in Aruba, Curaçao and Bonaire. Any newly introduced species has to be able to survive such dry periods in order to establish itself on the islands permanently.

Bufo marinus survives in Aruba by burrowing under drip systems and potted plants in gardens and under the outlets of air conditioners (van Buurt, 2005). In

Aruba the tankis almost always contain water, and fall dry only at the end of very long dry periods. For some species this may facilitate survival, even though Aruba is even more arid than Curaçao and Bonaire. The corn snake arrived in Curaçao during the 2001-2002 dry period. It was found in gardens and in small plots of land irrigated for horticulture, where it stood out so that many were captured and killed. It is possible that survival of the corn snake would have been much better in years with a good rainy season, which would have enabled it to spread out and find refuge in areas with less human presence, where mortality would likely be lower. The extended drought may thus have had an important role in our being able to eradicate the corn snake.

Acknowledgements. Paul Ch. Hoetjes, Eric Newton, Facundo Franken and Dr. Howard K. Reinert provided comments and information.

References

Barmes, T., Eckert, K.L., Sybesma, J. (1993): Sea Turtle Recovery Action Plan for Aruba. United Nations Environment Programme. Caribbean Environment Programme Technical Report No. 25.

Buurt, G. van (1999): 500 jaar natuur op Curaçao, verleden, heden en toekomst. In: Veranderend Curaçao. Pp. 87-121, Stichting Libri Antilliani, Bloemendaal, Nederland.

Buurt, G. van (2001): De Amfibieën en Reptielen van Aruba, Curaçao and Bonaire. Uitgegeven door de schrijver, Curaçao.

Buurt, G. van (2005): Field Guide to the Amphibians and Reptiles of Aruba, Curaçao and Bonaire. Edition Chimaira, Frankfurt am Main.

Debrot, A.O., Pors, L.P.J.J. (1995): Sea turtle nesting activity on northeast coast beaches of Curaçao, 1993. Carib. J. Sci. **31**: 333-338.

Debrot, A.O., Esteban, N., Le Saco, R., Caballero, A., Hoetjes, P.C. (2005): New sea turtle nesting records for the Netherlands Antilles provide impetus to conservation action. Carib. J. Sci.. **41**: 334-339.

Fish, M.R., Côté, I.M., Gill, J.A., Jones, A.P., Renshoff, S., Watkinson, A.R. (2005): Predicting the impact of sea-level rise on Caribbean sea turtle nesting habitat. Conserv. Biol. **19**: 482-491.

Hartog, J. (1957): Bonaire van Indianen tot Toeristen. De Wit Inc publishers, Aruba.

Hartog, J. (1961): Aruba, Past and Present; From the Time of the Indians Until Today. De Wit Inc publishers, Aruba.

Hartog, J. (1968): Curaçao: From Colonial Dependence to Autonomy. De Wit Inc publishers, Aruba.

Haviser, J.B. (1987): Amerindian Cultural Geography on Curaçao. Dissertation, Leiden University.

Haviser, J.B. (1991): The First Bonaireans. Reports of the A.A.I.N.A. no.10, Curaçao.

Hendrikse, N. (2005): Hollands Glorie in de Kolonie. Omni Media, Curaçao.

Hodge, K.V.D., Censky, E.J., Powell, R. (2003): The reptiles and amphibians of Anguilla, British West Indies. Anguilla Naional Trust.

IUCN (2004): 2004 IUCN Red List of Endangered Species. Gland, Switzerland. Accessed online at http://www.iucnredlist.org/.

Lammarée, L. (1970): Lizards of the genus *Cnemidophorus* from the Leeward group and the adjacent mainland of South America. Studies on the Fauna of Curaçao, and other Caribbean Islands. **34**: 46-72.

Marken Lichtenbelt, W.D. van, Albers, K.B. (1993): Reproductive adaptations of the green iguana on a semiarid island. Copeia **1993**: 790-798.

Marken Lichtenbelt, W.D. van, Wesselingh, R.A., Vogel, J.T., Albers, K.B. (1993): Energy budgets in free-living green iguanas in a seasonal enviromnent. Ecology **74**: 1157-1172.

Meshaka, W.E., Jr. (2001): The Cuban Treefrog in Florida: Life History of a Successful Colonizing Species. University Press of Florida, Gainesville.

Perry, G., Pierce, J. Griffin, D., van Buurt, G., Lazell, J. (2003): Corn snake distribution, *Elaphe guttata guttata*. Herpetol. Rev. **34**: 264-265.

Powell, R., Henderson, R.W., Parmerlee, J.S., Jr. (2005): The reptiles and amphibians of the Dutch Caribbean (St. Eustatius, Saba, and St. Maarten). St Eustatius National Parks Foundation (STENAPA).

Quick, J., Reinert, H.K., de Cuba, E.R., Odum, R.A. (2005): Recent occurrence and dietary habits of *Boa constrictor* on Aruba, Dutch West Indies. J. Herpetol. **39**: 304-307.

Reinert, H.K., Bushar, L.M., Odum, R.A. (1995): Recommendations for the conservation and management of the Aruba Island rattlesnake (*Crotalus unicolor*). In: Aruba Island Rattlesnake Conservation Action Plan. The Toledo Zoological Society, Toledo, U.S.A.

Reinert, H.K., Bushar, L.M., Rocco, G.I., Goode, M., Odum, R.A. (2002): A revision of the distribution of the Aruba Island rattlesnake, *Crotalus unicolor*, on Aruba, Dutch West Indies. Carib. J. Sci. **38**: 126-128.

Renkema, W.E. (1981): Het Curaçaose Plantagebedrijf in de negentiende eeuw. Zutphen, De Walburg Pers.

Stienstra, P. (1991): Sedimentary petrology, origin and mining history of the phosphate rocks of Klein Curaçao, Curaçao, and Aruba, Netherlands West Indies. Uitgaven Natuurwetenschappelijke Studiekring voor het Caraïbisch gebied: 130, Amsterdam.

Sybesma, J. (1992): WIDECAST Sea Turtle Recovery Action Plan for the Netherlands Antilles (Karen L. Eckert, Editor). CEP technical Report no.11, UNEP Caribbean Environment Programma, Kingston, Jamaica, 63 p.

Sybesma, J., Hoetjes, P.C. (1992): First record of the olive ridley and of nesting by the loggerhead turtle in Curaçao. Carib. J. Sci. **28**: 103-104.

Wagenaar Hummelinck, P. (1940): Studies on the Fauna of Curaçao, Aruba, Bonaire and the Venezuelan Islands. Dissertatie Utrecht.

Accepted: July 17, 2006 (AH).
Reprinted from Applied Herpetology **3**: 307-321 (2006).

Addendum

The Cuban treefrog now breeds in Curaçao:

Buurt, G. van (2007): Breeding population of *Osteopilus septentrionalis* in Curaçao. Appl. Herpetol. **4**: 390-391.

Status and conservation of the reptiles and amphibians of the Bermuda islands

Jamie P. Bacon[1,2], Jennifer A. Gray[3], Lisa Kitson[1]

[1] Bermuda Zoological Society, Flatts FL 04, Bermuda
[2] Corresponding author; email: jbacon@ibl.bm
[3] Bermuda Government Department of Conservation Services, Flatts FL 04, Bermuda

Abstract. Bermuda's herpetofauna includes three species of amphibians, one fossil tortoise, two species of freshwater turtles, five species of marine turtles, and four species of lizards. The amphibians *Eleutherodactylus johnstonei*, *E. gossei* and *Bufo marinus* were all introduced in the late 1880s. Amphibian population declines, including the possible extirpation of *E. gossei*, prompted the initiation in 1995 of an on-going investigation. Research into the high deformity rates in *B. marinus* has indicated that survival and development of larvae are affected by contaminants in a number of ponds and by the transgenerational transfer of accumulated contaminants. Of the two emydid turtles in Bermuda, *Malaclemys terrapin* may be native and its population characteristics are being studied; *Trachemys scripta elegans* is considered invasive and efforts are underway to remove its populations from the wild. The sizeable resident *Chelonia mydas* population has been the focus of a mark recapture study since 1968. Results indicate that Bermuda is currently an important developmental habitat for green turtles originating from at least four different nesting beaches in the Caribbean. Immature *Eretmochelys imbricata* also reside on the Bermuda Platform and genetics studies suggest that multiple Caribbean genotypes are represented in Bermuda's hawksbill population. *Caretta caretta* do not appear to be regular inhabitants, but two known loggerhead nesting events have recently occurred (in 1990 and 2005) and post-hatchling loggerheads regularly strand after winter storms. *Dermochelys coriacea* are only occasionally seen and the last record for a live *Lepidochelys kempi* in Bermuda occurred in 1949. Three of the lizard species are introduced *Anolis*; *A. grahami grahami*, *A. leachii*, and *A. extremus*. Their populations appear stable and they are presently not being studied. The fourth lizard, the Bermuda skink *Eumeces longirostris*, is Bermuda's only endemic terrestrial vertebrate. It is classified as Critically Endangered on the IUCN Red List and is protected under the Protected Species Act (2003); much research has been undertaken recently to aid the development of effective conservation management plans for this species.

Key words: Bermuda; *Bufo marinus*; *Caretta caretta*; *Chelonia mydas*; chemical stressors; deformities; *Eleutherodactylus gossei*; *Eretmochelys imbricata*; *Eumeces longirostris*; *Malaclemys terrapin*.

Introduction

Bermuda is an isolated 5,560 ha chain of limestone islands on a 150,000 ha seamount located near 32°N and 64°W in the western North Atlantic. Situated some 960 km ESE of Cape Hatteras, North Carolina, Bermuda consists of a crescent-shaped chain of more than 360 low-lying islands that are closely linked. A shallow shelf consisting of coral reefs, shallow lagoons and seagrass meadows surrounds the islands and makes up the Bermuda Platform (fig. 1).

The Bermuda islands are positioned within the north-western sector of the Sargasso Sea, a vast area of weak and inconsistent currents whose surface is dotted with mats of Sargassum algae. The Sargasso Sea offers a unique refuge to a host of open ocean species, including sea turtles. Driven by the Gulf Stream Current from the northwest and the Canaries Counter Current from the southeast, the Sargasso Sea turns slowly clockwise. The Gulf Stream passes Bermuda to the west with great influence as eddies and gyres reach Bermuda's shores and deliver warm water along with elements of the fauna and flora from the Caribbean and the east coast of North America.

While ocean surface temperatures range from 18°C in January to 28°C in August, the water mass surrounding Bermuda between the depths of 200 and 500 m is consistently about 18°C. Inshore temperatures may vary from 15°C to 30°C. Rainfall is not highly seasonal with a mean actual accumulation of approximately 150 cm being distributed throughout the year. October is the wettest month with an average of 16 cm, and April the driest at 10 cm. Temperatures show marked seasonality with mean monthly air temperatures ranging from 18.5°C in February to 29.6°C in August.

Seven of Bermuda's largest islands are connected by bridges and comprise what is considered to be 'mainland Bermuda'. The available land area (4,650 ha) is divided into nine parishes (fig. 1). Bermuda's topography is dominated by low rolling hills of poorly fused limestone and fertile depressions.

A number of ponds are scattered throughout Bermuda, but the majority are either fully marine or brackish and many are man-made. The island's few freshwater wetlands, estimated at 127 ha in the early 1600s, totaled only 58 ha in 1980 due to drainage for agriculture or mosquito control and through being used as landfill sites for waste disposal (Thomas, 2004). Presently, these freshwater habitats (some of which temporarily turn slightly brackish in the summer or fall) include one swamp forest, two marshes, two natural ponds and eight excavated ponds, some of which are located in former landfill sites. Additionally, there are a number of lined golf course ponds which provide fresh water habitats for terrapins and toads.

Currently, more than 50% of Bermuda's land area is used for housing and over 75% of Bermuda is considered developed (Thomas, 2004). With a population of over 61,000 and a population density of 1,145 people per km^2, Bermuda is one of the most densely populated oceanic islands in the world (Anderson et al., 2001). As a result, Bermuda's natural environment is at risk from chemical contamination caused by a variety of sources including illegal dumping, emissions from vehicles,

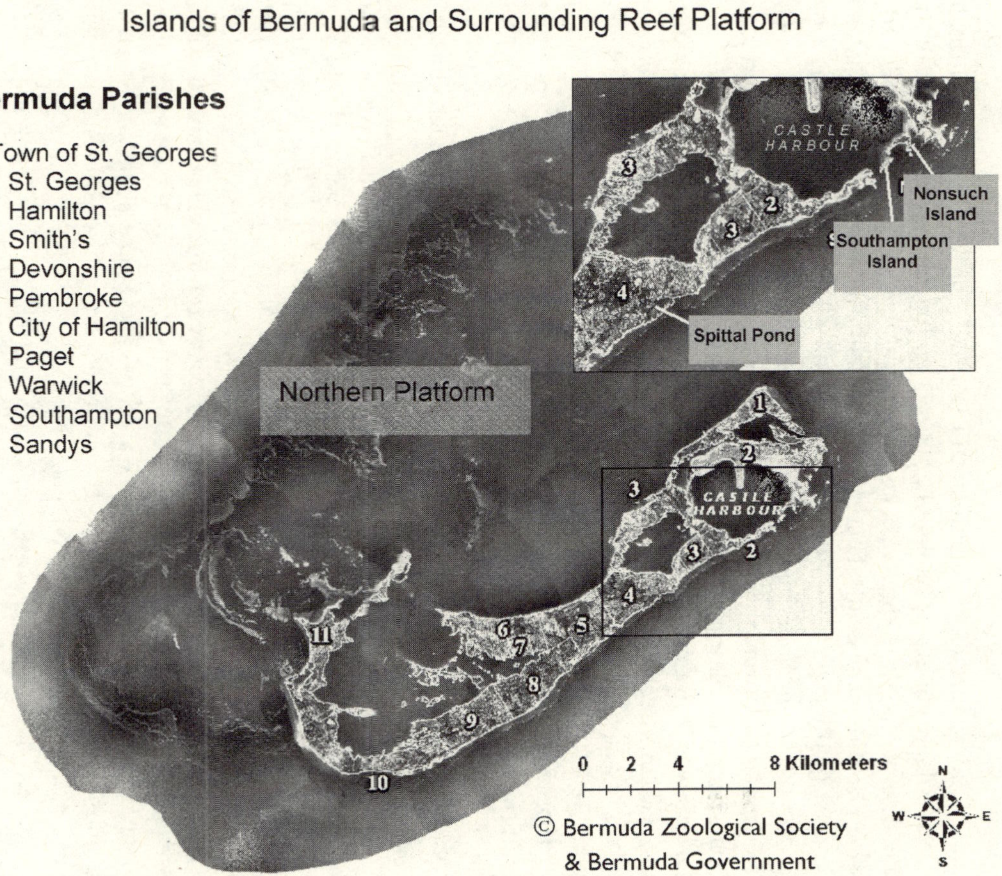

Figure 1. Map of Bermuda showing parishes, northern reef platform and key islands.

Bermuda's incinerator and electrical power plant, run-off from roadways and agricultural fields, and the leaching of contaminants from landfills, cesspits and deep-sealed bore holes.

The Amphibians

There are no native species of amphibians in Bermuda and this has been attributed to the archipelago's extreme isolation, recent geological origin and small size (Wingate, 1965). However, three species of West Indian anurans, all introduced in the late 1800s, did successfully become established on Bermuda's larger islands. These consist of the cane toad, *Bufo marinus*, and two species of whistling frogs, *Eleutherodactylus johnstonei* and *E. gossei* (table 1). As all three species were introduced to the island and their populations are not threatened globally, Bermuda's amphibian species are not protected under local conservation legislation.

B. marinus was deliberately introduced to Devonshire parish in 1885 in an effort to control garden insects (Wingate, 1965). In 1917, it was reported that *B. marinus* initially underwent a population 'explosion' which was followed by a decline to a stable level (Pope, 1917). When their status was reexamined by Wingate between 1956 and 1963, it was found that cane toads were 'common and universally distributed' on all of Bermuda's large inhabited islands (Wingate, 1965). Although the cane toad is recognized as a potentially damaging invasive species, there is currently no evidence to indicate that its presence poses an ecological threat in Bermuda.

The exact date of *E. johnstonei*'s introduction is unknown, but it was reported that they existed in very small numbers in Pembroke parish before 1880, when a pair from the Lesser Antilles was deliberately introduced in the same parish (Wingate, 1965). Based on the known range of *E. johnstonei* before 1880, it was hypothesized that the original population also arrived from the Lesser Antilles. By 1916, *E. johnstonei*'s range had expanded eastward through Hamilton parish and westward into Paget parish (Wingate, 1965). Its range continued to expand in subsequent years such that by 1963, it was considered very abundant on all the major islands excepting St. David's and even existed on some of Bermuda's smaller islets through man's introduction (Wingate, 1965).

It is believed that *E. gossei* was accidentally introduced into Pembroke, Paget, or Devonshire parish in vegetation imported from Jamaica in the 1890s (Wingate, 1965). By 1916, its range was still limited to Paget and Pembroke parishes (Wingate, 1965). Surveys in 1958 and 1963 revealed that its range, while including Warwick and Devonshire parishes, had become static, probably due to the fact that the more successful *E. johnstonei* had achieved an island-wide distribution (Wingate, 1965). Though there were also two separate colonies of *E. gossei* residing on large farms in Southampton, it was suggested that these populations were the result of individuals or eggs having been transported to these locations in manure (Wingate, 1965).

Table 1. Origin, distribution and conservation status of Bermuda's amphibian and reptile species.

Common name	Latin name	Ecological status	Date and origin	Distribution	Conservation status
Common whistling frog	*Eleutherodactylus johnstonei*	Introduced	L. 1800's Lesser Antilles	Island wide	Not evaluated
Whistling frog	*Eleutherodactylus gossei*	Introduced	L. 1800's Jamaica	Unknown	Locally extirpated
Cane toad	*Bufo marinus*	Introduced	1885 Guyana	Island wide	Not evaluated
Diamondback terrapin	*Malaclemys terrapin*	Native	N. America	In a few isolated ponds	Vulnerable (VU, D1)
Red-eared slider	*Trachemys scripta*	Introduced	L. 1900's N. America	Island wide in ponds	Invasive
Green turtle	*Chelonia mydas*	Native	–	Inner bays and harbours and north platform reefs	Endangered (EN, A1 b,d)
Loggerhead turtle	*Caretta caretta*	Native	–	Sargassum rafts and occasionally near wrecks	Endangered (EN, A1 a,b,d)
Hawksbill turtle	*Eretmochelys imbricata*	Native	–	Reefs island wide	Critically endangered (CR, A1 b,d)
Atlantic Ridley turtle	*Lepidochelys kempi*	Rare visitor	–	–	Endangered (EN, A1 b,d)
Leatherback turtle	*Dermochelys coriacea*	Migratory visitor	–	Offshore	Endangered (EN, A1 b,d)
Bermuda skink	*Eumeces longirostris*	Endemic	Bermuda	Island wide in isolated pockets	Critically endangered (CR, B1, B2 b,c,d,e)
Jamaican anole	*Anolis grahami*	Introduced	1905 Jamaica	Island wide	Not evaluated
Antiguan anole	*Anolis leachii*	Introduced	1940 Antigua	Island wide	Not evaluated
Barbados anole	*Anolis extremus*	Introduced	1940 Barbados	Restricted mainly to the western parishes	Not evaluated

Despite the narrow range identified in the 1963 survey, there was no indication that *E. gossei* populations were in decline at that time.

By the mid 1990s, however, it became apparent that the population of *E. gossei* was declining, and indeed might have been extirpated from Bermuda since no specimens of this species had been observed since 1994. At about this time, there was also concern that the populations of *E. johnstonei* and *B. marinus* were declining (Royal Gazette, 1993). In response to these concerns, the Bermuda Amphibian Project was initiated in 1995.

Local residents and scientists suggested that habitat destruction and chemical pollution had contributed to the apparent declines in the island's amphibian populations (Wingate, pers. comm.; Linzey et al., 2003). Therefore, many of the initial investigations focused on identifying the environmental stressors potentially affecting Bermuda's amphibians. To do this, soil samples, water samples, and amphibian tissue samples were collected from 15 study sites between 1995 and 1999 and analysed for pesticides and heavy metals. The analyses revealed that p,p′-dichlorodiphenyldichloro-ethylene (DDE) was found in soil in concentrations ranging from <0.1 ppm to 1.2 ppm at the 10 sites where soil was sampled. DDE was also found in the livers and fat bodies of toads and whistling frogs from all six sites where specimens were collected as well as in invertebrate prey items from the one site where invertebrates were collected indicating possible transport through the food chain (Linzey et al., 2003). Although use of dichlorodiphenyltrichloroethylene (DDT) in Bermuda ceased in 1972, its metabolite, DDE, is apparently still ubiquitous across the main islands. Additionally, pesticide residues in soil samples included DDT at eight sites, kelthane at eight sites, dieldrin at five sites, and polychlorinated biphenyls (PCBs) as Arochlor 1254 and Arochlor 1260 at seven sites (Linzey et al., 2003). Furthermore, the data suggested that heavy metals from the environment were another potential stressor since analyses of toad livers revealed significant concentrations of cadmium, chromium, copper and zinc (Linzey et al., 2003).

Further studies strongly suggested that Bermuda's whistling frogs and toads were exhibiting effects caused by exposure to environmental stressors. Both species were found to harbour heavy parasite loads and both were infected with multiple species of nematodes and trematodes (Linzey et al., 1998a; Linzey et al., 1998b). In addition, histopathological and lymphocyte proliferation studies indicated that immune function was being suppressed in both species (Linzey et al., 2003). However, the sample sizes used in the immune function studies were small and this is being investigated further.

More recent investigations have focused on the disturbingly high incidence of abnormalities in Bermuda's cane toad populations (table 2) and the possible implications for environmental health. Surveys of adults and juvenile toads and metamorphic toads have shown that the problem is persistent and widespread. Abnormal toads were found in a variety of habitats in all nine parishes (Bacon et al., 2006). While most abnormalities in both age classes involved skeletal malformations of the

Table 2. Overall field abnormality frequencies in adult/juvenile and newly metamorphosed *Bufo marinus*. This table is updated from table 1 of Bacon et al. (2006) by addition of data for 2004 and 2005.

Adult/juvenile *B. marinus*[1]					Newly metamorphosed *B. marinus*				
Year	Number of sites[2]	*n*	Abnormal (%)	Range per survey[3]	Year	Number of sites[2]	*n*	Abnormal (%)	Range per cohort[3]
1999-2000	27	726	19.1	0-29%	2000	18	2,223	15.9	0-47%
2001	10	545	26.4	12-38%	2001	17	3,687	19.4	0-61%
2002	11	521	30.1	6-43%	2002	24	3,520	21.4	0-81%
2003	11	682	27.9	15-55%	2003	18	1,952	24.0	0-64%
2004	13	894	30.2	13-48%	2004	16	2,204	20.4	0-46%
2005	10	718	28.5	18-49%	2005	16	1,888	20.5	2-49%

[1] Includes reproductively mature adults and juvenile specimens.
[2] Number of sites surveyed per year.
[3] Range of abnormality frequencies found during the given survey year.

hind limbs, a variety of spinal, pelvic, and facial (predominantly eye) abnormalities were also observed (Bacon et al., 2006). In addition, annual metamorph abnormality rates at particular ponds were as high as 58% ($n = 71$ from four collections) and abnormality rates for particular cohorts were as high as 81% ($n = 26$ for this cohort; mean cohort size c. 67 individuals) (Bacon et al., 2006). Data from 2000-2003 revealed that breeding sites in public areas (natural or excavated ponds in parks and nature reserves, some of which were former landfill sites, or lined ponds on golf courses) had significantly higher abnormality frequencies than sites (lined or cement ponds) in backyard settings (Bacon et al., 2006).

The possibility that parasites were causing the *B. marinus* abnormalities was investigated in 2001 and 2002. However, no encysted *Ribeiroia* metacercariae were found in 80 malformed metamorphs collected from four sites with high abnormality rates, indicating that these parasites were not responsible for the abnormalities observed (Bacon et al., 2006). Subsequently, investigations were conducted to ascertain if chemicals were responsible for the abnormalities observed. To accomplish this, a number of water and sediment samples and embryos were collected beginning in 2002 from five ponds with high abnormality rates and two reference ponds for use in frog embryo-larval teratogenesis assays. Results from these assays revealed that water and sediment extracts from each of the five affected ponds induced severe abnormalities in the developing larvae of three amphibian species including *B. marinus* (Bacon et al., 2006; Fort et al., 2006a). Further studies showed that developmentally toxic sediment samples contained elevated levels of petroleum hydrocarbons, metals (aluminum, arsenic, tin, cadmium, chromium, copper, iron, lead, mercury, manganese, nickel, and zinc) and ammonia (Fort et al., 2006a).

More recent investigations have indicated that the levels of petroleum hydrocarbons and metals in sediments from the affected ponds were capable of inducing developmental malformations in *B. marinus* independently of each other (Fort et al., 2006a). However, joint mixture interaction studies also suggested that the two classes of pollutants act synergistically when both are present, and these findings have significant environmental health implications as all five of the ponds tested contained both classes of pollutants (Fort et al., 2006a). Larval exposure to the identified contaminants was confirmed through tissue residue analyses (Fort et al., 2006a), and chronic exposure studies have also indicated an association between the amounts of contaminants absorbed and the frequencies of developmental malformations observed in *B. marinus* metamorphs (Fort et al., 2006a). More recently, cross-over exposure studies, in which *B. marinus* embryos from contaminated sites were raised in reference site media and vice versa, showed that there was also a marked maternal effect on larval development and survival. These studies suggested that metals and petroleum hydrocarbons were being transferred from the mother to her eggs during oogenesis and oocyte maturation, and that this transfer of contaminants had a marked impact on larval development and survival (Fort et al., 2006b). In summary, investigations to date indicate that survival and development of *B. mar-*

inus larvae in Bermuda are being affected both by contaminants found in a number of its ponds and by transgenerational transfer of accumulated contaminants.

The focus of current research is to complete the investigations and analyses required in order to determine and communicate the probable risks that the identified environmental contaminants pose to Bermuda's amphibians and potentially other species including humans. These data will be presented to Bermuda's government and environmental NGOs for use in the development of management plans for amphibians and affected sites and, if appropriate, remediation plans for affected sites.

The Turtles

Records for members of the Testudines in Bermuda exist for one fossil land tortoise, two freshwater turtles and five marine turtle species (table 1). The land tortoise, *Hesperotestudo bermudae,* was described from the Pleistocene of Bermuda. This single fossil was discovered in 1991 during the excavation of a fossilized sand dune and is thought to be some 300,000 years old (Meylan and Sterrer, 2000). *Hesperotestudo* has a long record in North America and these authors hypothesized that it rafted to Bermuda, perhaps using the Gulf Stream.

The brackish and freshwater ponds of Bermuda support populations of two emydid turtles; the diamondback terrapin, *Malaclemys terrapin,* and red-eared slider, *Trachemys scripta elegans*. There is a single record of *Malaclemys* from a cave in Bermuda (Thomas, 2004), but the bone does not appear to be fossilized and was found in association with pig bones (*Sus scrofa*). Pigs were likely introduced to Bermuda by Spanish sailors prior to 1535. Presently, two natural brackish water ponds and a few small adjacent golf course ponds support a population of this species. In these ponds, *Malaclemys* has been observed to forage among the submerged roots of red mangroves, *Rhizophora mangle* (Thomas, 2004). They are also known to lay their eggs in the sand bunkers of a local golf course in Hamilton Parish (Davenport et al., 2005). Preliminary study indicates a small but stable population and ecologists for the Bermuda Biodiversity Project are currently assessing the population characteristics, which will include total estimated population, density, ecological role, and genetic identity. There is no recorded evidence that *Malaclemys* were ever imported into Bermuda and the fact that they are well suited to long distance travel at sea raises the possibility that the species could be native (Davenport et al., 2005). If further study suggests that *Malaclemys* is native, this species would likely be given serious consideration as an addition to the Protected Species Act 2003.

The red-eared slider (*Trachemys scripta elegans*) is considered an invasive species in Bermuda, having been introduced to the island through the pet trade. The full extent of its impact on freshwater ecosystems in Bermuda has not been extensively studied but this species is now known to be established in numerous countries around the world. Recent research provides evidence that introduced *T. s. elegans* negatively impacts at least some native species in Europe (e.g., Cadi and Joly, 2003, 2004). Recent surveys in Bermuda have found it in almost every freshwater and wetland ecosystem and even a few brackish water ponds. Efforts are now underway to remove populations from the wild and raise awareness in the community regarding their potential impact.

Five species of marine turtles have been recorded from Bermuda with two, the green turtle, *Chelonia mydas,* and the loggerhead turtle, *Caretta caretta*, being the only species recorded nesting. The large breeding population of green turtle that was described in early histories of Bermuda was quickly decimated, through overharvest, by the first colonists and New World explorers, leading to one of the earliest pieces of conservation legislation, written in 1620, to protect young sea turtles (Garman, 1884). This early attempt at green turtle conservation was unsuccessful in maintaining the nesting population and there has been no reproduction of green turtles on the islands of Bermuda since the early 1900s (Babcock, 1937; Gray, unpublished data). However, in an effort to re-establish a breeding population of *C. mydas*, more than 25,000 green turtle eggs were collected from beaches in Costa Rica and Surinam and reburied on local beaches between 1967 and 1977. Approximately 16,000 hatchlings were produced during this restocking experiment (Department of Agriculture and Fisheries Monthly Bulletin, 1972). Due to the late incubation period and cool sand temperatures it is likely that these hatchlings had a male-biased sex ratio. No green turtle nesting has been recorded in Bermuda since the restocking project was carried out.

At the present time significant numbers of juvenile green turtles and a smaller number of hawksbill turtles from nesting populations in the Caribbean region forage and reside in Bermuda. The shallow reefs and seagrass meadows of the northern Bermuda Platform (fig. 1) provide excellent grazing areas for the large numbers of resident green turtles. These animals are the focus of a mark and recapture study, initiated in 1968, by the Bermuda Turtle Project (BTP) (Meylan et al., 1994; Gray et al., 1998). Turtles are captured for study using a 2000 ft entrapment net set at 40 sites around the island (fig. 2, c & d). Data are collected from all turtles captured by the BTP using a standardized protocol (Meylan et al., 1992-2003). As of December 2005, data from approximately 2,500 green turtles have been collected and more than eight hundred recaptures have been made. These observations provide extensive data on population structure and trends, genetic identity, sex ratios, growth rates, site fidelity and migratory patterns.

BTP studies indicate that the very small, pelagic size class and mature adults are absent from the Bermuda Platform. This supports the hypothesis that Bermuda is a developmental habitat for green turtles (Meylan et al., 1994; Meylan and Meylan, 1998). Green turtles captured by the BTP have varied in minimum straight carapace length at first observation from approximately 22 to 81 cm and in weight from 1 to 86 kg (Meylan et al., in prep.). Although a small number of the green turtles captured are larger than the minimum size at sexual maturity, laparoscopy of a sample of more than 125 individuals suggests that none are mature (Meylan et al., 1994; Meylan

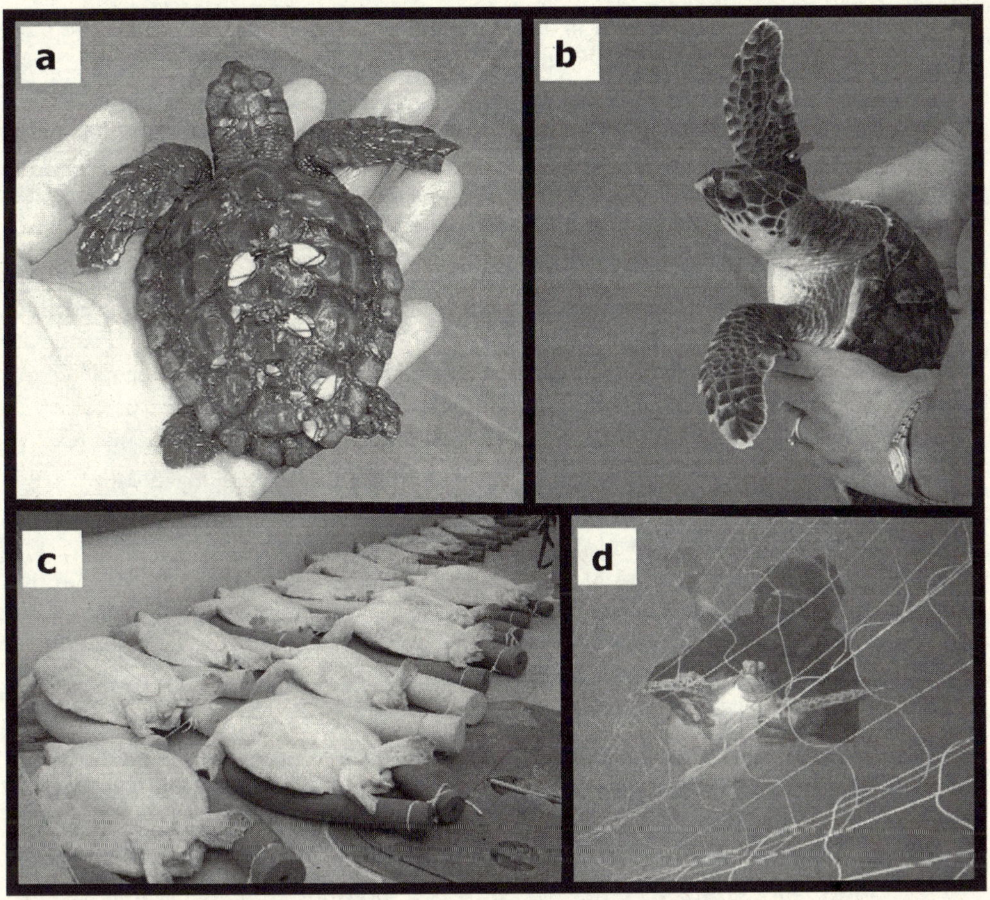

Figure 2. Sea turtles: (a) juvenile loggerhead; (b) juvenile hawksbill; (c) green turtles on research vessel; (d) green turtle capture in net.

et al., in prep.). Among the green turtles observed through the Bermuda Aquarium stranding network, 8.4% are smaller than the minimum size captured on the seagrass flats by the mark and recapture study. This would suggest that Bermuda receives a small number of pelagic phase green turtles that become sick or injured in the oceanic environment.

Bermuda's green turtles have been found to maintain grazing plots on seagrass beds (Vierros et al., 2002). Furthermore, the capture of green turtles during every month of the year suggests that they are year-round residents of the platform, staying in Bermuda's waters as long as 14 years or more (Meylan et al., in prep.).

Blood samples are taken from each captured turtle to determine gender and genetic affinities through sequencing of the mitochondrial control region. Data indicate that at least four different nesting beach populations contribute to the group of green turtles that forage on the Bermuda Platform (Engstrom et al., 1998; Vierros et al., 2002).

Upon reaching a shell length of approximately 65-70 cm, green turtles depart from Bermuda and migrate to distant foraging grounds where they complete their development and become sexually mature (Meylan et al., in prep.). External tags allow researchers to determine the locations of these distant foraging grounds. To date, over 90 green turtles and one hawksbill tagged by the BTP in Bermuda have been recaptured in other countries bordering the Caribbean. Most green turtle recaptures have been made in Nicaragua, reflecting travel of about 2500 km straight-line distance. The tag return data suggest that the turtles take up residence on the extensive shallow grass beds off the coast of this country. This region is the primary feeding ground for mature adult green turtles in the western Caribbean. From this area, mature turtles undertake reproductive migrations to the nesting beach where they were born, completing a long and complex life cycle (Carr et al., 1978).

Hawksbills (*Eretmochelys imbricata*) are found on the reefs from the NW to NE of the islands and occasionally on the south shore reefs. They are only very rarely caught in the entrapment net used to capture green turtles, but a small sample has been collected by a variety of methods (Meylan et al., 2004). Since 2000, the BTP has dedicated annual sampling effort to swimming transects of suitable habitat with teams of snorkelers, capturing hawksbills by hand. This has proven to be a successful technique, but the apparent low density of the species precludes a large sample size for study. The majority of the hawksbill data from Bermuda have been collected from stranded animals reported to the Bermuda Aquarium Wildlife Rehabilitation Centre and through animals captured by cooperating recreational divers licensed to take lobsters. A data set for 136 hawksbills (caught by hand or in nets, or observed as strandings) shows a straight carapace length range from 5.3 cm to 75.7 cm with the smallest individuals being from the stranding network (fig. 2b).

Three of the hawksbills examined approached the minimum size at maturity for hawksbills in the Atlantic. However, necropsies of stranded animals have yet to reveal any sexually mature individuals. Furthermore, no nesting by hawksbills has been documented in Bermuda. Thus, the hawksbills known from Bermuda appear to represent residents from a developmental habitat on the Bermuda Platform and stranded individuals from the pelagic life history stage (Meylan et al., 2004).

The mitochondrial control region has been sequenced for a sample of Bermuda hawksbills, and multiple Caribbean genotypes have been detected. These include haplotypes known in 2002 from nesting beaches on Cuba, U.S. Virgin Islands, Mexico and Costa Rica (Meylan et al., 2004). However, reanalysis of the growing Bermuda data set in light of the increasing knowledge of hawksbill genetics will be required to corroborate these sites as source populations for the hawksbills in Bermuda.

The loggerhead, *Caretta caretta*, is not known to regularly inhabit the waters of the Bermuda Platform nor is it the main target of research. However, during winter months, juvenile, post-hatchling *Caretta* strand on Bermuda shores, typically washed in by winter storms (fig. 2a). They vary in size from 6.26 cm to 74.5 cm straight carapace length with more than 75% being less than 20 cm. Live, stranded loggerheads are cared for in the Bermuda Aquarium Museum and Zoo's Wildlife Rehabilitation Center and then tagged with BTP tags before release. There are anecdotal reports of large *Caretta* on or near the outer reefs and wrecks, but such records have not been confirmed with size or species verification (BTP sighting log). In 1990, the first nesting by *Caretta* in Bermuda was recorded (Gray, 1990) and this was followed by a second loggerhead nesting event in 2005 (Gray, in prep.). While sea turtle nesting in Bermuda is extremely rare, with only two nests recorded in nearly two decades, there is nonetheless a recognized need to protect Bermuda's remaining habitat for any returning breeders.

Leatherback turtles, *Dermochelys coriacea*, are only occasionally seen offshore and there are only five records of strandings since 1967. Three of these were a result of entanglement in fish or lobster trap lines. Large leatherbacks are seen off the edge of the Bermuda Platform and very occasionally within the reef system, presumably passing the islands on their migration to or from other regions in the North Atlantic.

Two museum records from the 1940s exist for Kemp's Ridley turtle, *Lepidochelys kempi*, in Bermuda. It was also presumed that this species had visited the island on rare occasions previously (Babcock, 1937). Other than anecdotal records, however, there is only one live *Lepidochelys kempi* account on record for Bermuda of a specimen caught in 1949 and held by the Bermuda Aquarium (Mowbray and Caldwell, 1958). More recently, in April of 2006 an additional dead specimen of *Lepidochelys kempi* was collected by the Wildlife Rehabilitation Centre at the Bermuda Aquarium Museum and Zoo.

The degree of protective legislation for turtles in Bermuda has been progressive. The earliest known legislation protecting marine turtles was passed by the First Assembly in 1620, only eleven years after Bermuda was colonized. The act prohibited the taking of young turtles, protecting all turtles less than 18" breadth or diameter. In 1937, there was a prohibition of taking turtles under 20 lbs. The Board of Trade (Fisheries) regulations (1947) enforced a restriction on the take of any turtle of a weight smaller than forty pounds. An order made under the Fisheries Act (1972) implemented a moratorium on the take of all turtles for a five year period. This moratorium was never lifted but rather was replaced with the Fisheries Protected Species Order of 1978. To this end, all sea turtles in Bermuda have been completely protected from direct take since this act commenced on April 1, 1973. Sea turtles are further protected today under the Protected Species Act of 2003.

Sea turtles in Bermuda are a shared regional resource and as such are threatened by factors in other parts of the region that are not possible for us to track. Such threats may include, but are not limited to, harvest by humans, incidental catch, loss of habitat and natural predation. Threats to sea turtles on the Bermuda Platform are monitored by the stranding network which reports that 43% of local strandings are directly attributable to anthropogenic causes (Gray et al., 2006). Incidental catch and entanglements (entanglements in nets, fishing line or kite string; ingestion of fishing hooks) caused 22% of strandings in which a probable cause of death could be determined. An additional 18% of strandings appeared to involve boat collisions or propeller injuries, although some boat interactions may have occurred post mortem and may not have been the cause of the stranding. Death due to ingestion of plastic contributed to 3% of strandings. Necropsy results also reveal the presence of ingested plastics in the stomach contents and intestinal tracts of animals where death was caused by another factor. The negative effect of these contributing factors is likely to be underestimated, as partially dismembered carcasses that are too decomposed to necropsy may also be the result of human impacts. In addition to the threats of entanglement, boat collision, and ingestion of plastics, mortality of sea turtles stranded in Bermuda can be attributed to but not limited to gut impaction, drowning, toxaemia, foreign body ingestion, parasite burden, septicaemia, lung infection, clostridium burden, degenerative tissue disease and emaciation (Gray et al., 2006). A solution to the problems of incidental catch, ingestion of marine debris and entanglement will require international cooperation because in many cases the entangling and ingested materials do not have their origin in Bermuda.

Habitat loss as a result of a 30% decline in seagrass beds locally is an issue of great concern in Bermuda today. The potential effects to Bermuda turtles of the loss of nearly 500 hectares of critical marine habitat are of extreme concern, as is the lack of data that could aid in explaining this dramatic decline (Murdoch et al., in prep.).

The Skinks and *Anolis* Lizards

There are four species of extant terrestrial reptiles in Bermuda and all of them are lizards (table 1). Only one species, the endemic Bermuda skink, *Eumeces longirostris*, is native to the island while the other three, all *Anolis* species, are introduced.

The Jamaican anole, *Anolis grahami grahami*, is the most common of the three species of introduced lizards. It was introduced intentionally in 1905 to control the Mediterranean fruit fly, *Ceratitis capitata*. This lizard successfully colonised the mainland and most of the offshore islands. *A. grahami* is usually found on walls, trees and shrubs. It preys on a variety of insects including those introduced as biological pest controls, most notably a ladybird beetle, *Rhyzobius lophanthae*, that was introduced to combat infestations of the scale insect, *Carulaspis minima* on endemic cedar and palmetto trees (Thomas, 2004). This anole is also known to consume juvenile Bermuda skinks (Giffith and Wingate, 1994) and eggs of the native Eastern bluebird, *Sialia sialis* (Thomas, 2004). A bird from Trinidad, the Great Kisakadee, *Pitangus sulphuratus*, was introduced in 1957 in an unsuccessful attempt to control *A. grahami*. However, these birds also became predators of the Bermuda skink (Wingate, 1965; Thomas, 2004).

A second anole, *Anolis leachii*, native to Antigua and Barbuda, was accidentally introduced and first spotted in 1940 at the north end of Warwick parish. It is locally known as "the Warwick lizard" although the distribution of this species is now island wide. It is the largest lizard species in Bermuda (males reach up to 35 cm) and is dominant over the other anole species. *A. leachii* eats a wide variety of insects (Thomas, 2004) and Wingate (pers. comm.) hypothesizes that these anoles may also eat bluebird eggs and bluebird hatchlings from the nest because they are often found in vacant bluebird boxes and there are instances of egg and chick disappearances from boxes where this species is seen in the vicinity.

The third anole, *Anolis extremus* native to Barbados, was also accidentally introduced via ships at the west end of the island (the naval dockyard) around the same time as the introduction of *A. leachii.* This species is less common than the other two anoles and is largely restricted to the western parishes. It eats insects, spiders and woodlice (Thomas, 2004).

The Bermuda skink, also locally known as a "rock lizard", is the only endemic terrestrial vertebrate in Bermuda (fig. 3). Unless noted otherwise, information on this species is derived from a thesis by L. Kitson (in prep.). The Bermuda skink is a diurnal, ground-dwelling lizard that is thought to have evolved after an ancestral species colonised the islands via an oceanic journey (rafting) from the east coast of North America. The species differs widely from other members of the genus *Eumeces*. Heilprin (1889) suggested that the Bermuda skink's closest relative was *Eumeces fasciatus,* the common five-lined skink, from the south-eastern United States. Preliminary genetics studies suggest, however, that *E. longirostris* is more closely related to the western skink, (*Eumeces skiltonianus*) (Richmond, pers. comm.). This skink is common to parts of the Western United States and does not inhabit the east coast. It is likely that there is no extant common ancestor to the Bermuda skink and that this species evolved very early on during the radiation of the *Eumeces* genus (Richmond, pers. comm.).

Jones (1859), stated that the Bermuda skink was observed island wide particularly on walls and stones in cedar groves. However, 43 years later Verill (1902) reported that Bermuda skinks were rarely seen on the mainland and were prevalent only on Castle Island around the forts and cliffs. The difference between these two reports suggests that there was a dramatic decline of Bermuda skinks within a short period. However, the virtual absence of skinks in early records preceding the report made by Jones (1859) may imply that the skink was never particularly conspicuous. Wingate (1965) suggested that skinks were more subtly abundant on the mainland than previously thought, although in more recent years local residents have noticed a dramatic decline in numbers (Bermuda Biodiversity Project, unpublished).

E. longirostris is classified as critically endangered on the IUCN Red List (Conyers and Wingate, 1996). This species is protected locally under the Protected Species Act (2003) (table 1). The most significant threat to this species has undoubtedly been habitat loss due to the expanding area taken up by homes and gardens (now 50% of the total land area) and agriculture (17% of the total land area). However, declining skink numbers are also evident in nature reserves with restored vegetation and this has been attributed to predation by introduced species (Davenport et al., 2001).

(a)

(b)

Figure 3. (a) Juvenile (1 year old) Bermuda skink *Eumeces longirostris*. (b) Adult (4 year old) Bermuda skink. (Colour originals — see www.ahailey.f9.co.uk/appliedherpetology/cariherp.htm).

Efforts to understand the ecology of this species have been undertaken recently in order to develop effective conservation management plans. These include studies on population health and size (Raine, 1998; Wingate, 1998; Davenport et al., 2001; Glasspool and Outerbridge, 2005) species distribution, genetic differentiation of populations, reproductive biology, seasonal activity and prey availability.

In 1998, the Bermuda Zoological Society sent out a public questionnaire to all households. The results showed that there were possibly many isolated populations spanning the entire mainland. These reports were followed up by surveys using baited traps in areas where skink sightings had been reported. Skinks are now known to inhabit many rocky coastal areas throughout the mainland of Bermuda. However, they appear to be absent from the islands of St. George's and St. David's. Verill (1902) documented that there was a plague of rats in the 17th century and, in an attempt to control the rat population, the vegetation on these islands was repeatedly burned and domestic cats were deliberately shipped to Bermuda and introduced onto these islands. It is possible that skinks were eliminated from these islands (where humans first settled) as a result of this dramatic episode.

The greatest known abundance of skinks on Bermuda's mainland can be found within the Spittal Pond Nature Reserve (fig. 1). Raine (1998) estimated there were at least 124 individuals there although the entire rocky coastal area of the nature reserve was not surveyed during this study.

There are also substantial populations of skinks on several offshore islands, the largest of which is Southampton Island (fig. 1). Mark and recapture surveys on Southampton Island in 1997 showed that the population was healthy with adequate recruitment, which may have reached carrying capacity at approximately 400 adult animals (Davenport et al., 2001). Glasspool and Outerbridge (2005) later adjusted this figure to 534 in order to account for an assumed absence of mature females. Females brood eggs in early summer (Kitson, unpublished), the period when the original survey took place. In 2004, this island was surveyed again and estimated to contain 582 skinks (Glasspool and Outerbridge, 2005). Southampton Island appears to have the highest density of skinks in Bermuda and is regarded as the only safe haven for *E. longirostris* (Davenport et al., 2001) because introduced predators and anthropogenic threats are largely absent. However, in 2004 Glasspool and Outerbridge (2005) noted that the mutilation rate had increased since 1997. They suggested that the most likely cause was storm-induced stress, however they could not rule out increased avian predation.

Nonsuch Island is a protected nature reserve that is being restored in order to achieve a true representation of the prehistoric "native" environment of Bermuda. Detailed observations and surveys of *E. longirostris* have been carried out on this island since the 1970s (Wingate, pers. comm.). In 1997, a survey of the island's skink population indicated that it was ageing and failing. It is suggested that this decline resulted from predation by kiskadees, Jamaican anoles, and possibly cane toads. More than 50% of this population showed mutilation (damaged tails or digit loss). It is suspected that extinction of the Nonsuch Island population is inevitable without control of predators (Davenport et al., 2001).

Several studies have been carried out to investigate aspects of isolated skink populations. This work has added knowledge on environmental and genetic stress (Raine, 1998), population size, distribution and structuring (Davenport et al., 2001; Raine, 1998; Glasspool and Outerbridge, 2005). Raine (1998) found that skinks on Charles Island and Inner Pear Island and at Spittal Pond Nature Reserve were suffering from significant levels of environmental and genetic stress even though all three populations appeared healthy, with sufficient numbers of juveniles and adults.

Raine (1998) found significant morphological differentiation between geographically isolated populations and, in a study using microsatellite markers developed by Coughlan et al. (2004), significant genetic differences were found between skinks at Spittal Pond Nature Reserve and those on Southampton Island. In order to preserve the genetic diversity of *E. longirostris* and avoid outbreeding, it will be important not to mix the gene pools of geographically isolated populations in any future captive breeding and reintroduction attempts.

Recent research found that the skinks in Spittal Pond Nature Reserve are more genetically diverse than those on Southampton Island. Results from this study also suggest that the Spittal Pond population has suffered a dramatic decline recently or that there are barriers preventing intermixing within the population. Both populations have a healthy level of genetic variation compared to other species of island reptiles and endangered lizards.

Deliberate and accidental introductions are thought to have had dramatic effects on *E. longirostris* (Davenport et al., 2001; Raine, 1998; Wingate, 1998). The black rat, *Rattus rattus*, brown rat, *Rattus norvegicus* and domestic cat, *Felis domesticus*, have been resident on the island since the early human settlers. These species are considered threats to *E. longirostris*. There are currently thousands of feral cats in Bermuda (Glasspool, pers. comm.) and cases of domestic cats catching and killing skinks are frequently reported by their owners.

The great kiskadee, *Pitangus sulphuratus*, is an adept skink predator (Samuel, 1975) and is presently perceived to be the only bird that creates a major threat to the remaining skink population (Raine, 1998; Davenport et al., 2001). Yellow crown night herons, *Nyctanassa violacea*, were introduced between 1976-78 to replace an extinct native heron. Night herons are also known to eat skinks although their diet mostly consists of crabs (Davenport et al., 2001).

Griffith and Wingate (1994) observed a Jamaican anole eating a juvenile skink on Castle Island. Wingate (1998) suggested that, although *A. grahami* occupies a separate niche by being primarily arboreal, it could be a major predator of *E. longirostris* and the reason why skinks are more abundant on islands where these anoles are absent. Juvenile Jamaican anoles can frequently be observed foraging on the ground in skink habitats, and therefore almost certainly compete directly at times with hatchling and juvenile skinks (Edgar, pers. comm.). Cane toads, *Bufo marinus*, are considered a potential predator of Bermuda skinks (Davenport et al.,

2001). They could also compete with skinks for prey although they are nocturnal and not usually found in the same habitat.

Heilprin (1889) originally suggested that skinks were primarily insectivorous, in addition to preying upon isopods and amphipods. Wingate (pers. comm.) later observed that skinks also consumed carrion opportunistically and eggs from tropic bird, *Phaethon lepturus*, and cahow, *Pterodroma cahow*, nests. Davenport et al. (2001) discovered that skinks also eat prickly pear fruit, *Opuntia dilleni*. More recent studies, which examined fecal samples and gut contents from dissected specimens (incidental death), indicate that skinks prey upon a wide variety of arthropods including small crustaceans (isopods and amphipods), crickets, spiders, cockroaches, beetles and flies. Studies carried out in captivity show that skinks will eat the majority of terrestrial arthropods offered to them.

Prey availability surveys carried out in several sites show that out of all the arthropod species consumed by skinks, the most abundant are amphipods and isopods. These surveys were carried out in areas inhabited by skinks and in areas where they appeared to be absent (after pitfall surveys). It was revealed that isopods and amphipods were more available in rocky coastal areas where skinks were found compared to an upland hillside area where they appeared to be absent. It is suggested that this is one of the reasons why skinks are found mainly in rocky coastal areas.

It was found that *E. longirostris* has a life history similar to skinks of the same genus in North America. Adults court and mate in late April/early May and lay one clutch of eggs in late May/early June with hatchlings appearing in late July/early August. In captivity females brood their eggs for the entire incubation period. Studies in captivity show that Bermuda skinks have a mean clutch size of 4.5 ($n = 9$) and can lay up to 6 eggs. Incubation (in captivity) lasts an average of 36.3 days ($n = 6$). Captive skinks may reach reproductive maturity at two years of age and adult size after three years.

Bermuda skinks were originally thought to hibernate during the cooler winter months. However, recent studies now show that they remain active throughout the year although they are found (using baited traps) in greater abundance during the summer months.

Litter is considered to be a threat to the survival of this species (Davenport et al., 2001). Glass bottles are potentially lethal traps for skinks since they are unable to climb up steep smooth surfaces. Remains confirmed to be those of *E. longirostris* have been found inside bottles on the mainland (including in nature reserves) and on islets (Wingate, 1998; Wingate, pers. comm.).

Hurricanes occasionally affect Bermuda and cause the destruction of skink habitat. In 2004, Glasspool and Outerbridge (2005) noted that the 2003 cohort was virtually missing from Southampton Island. This cohort would have hatched shortly before Hurricane Fabian (direct hit, category 3) and it is suggested that this hurricane was primarily responsible for the lack of individuals in this size class. Pesticides and other chemicals in manicured areas may also threaten skinks. In one reported instance a dead skink was found in a flowerbed shortly after a pesticide had been used.

A Species Action Plan has been drawn up for the Bermuda skink with the help of the Herpetological Conservation Trust (U.K.) in order to ensure that this species is effectively managed (Edgar et al., 2006). Habitat restoration is currently the top priority for the conservation of this species and the replanting of native and endemic plants is an ongoing island-wide government initiative.

Acknowledgements. The Bermuda Turtle Project (BTP) was founded "to promote the conservation of marine turtles through research and education". Entering its 38th year in 2006, this is one of the longest-running projects of its kind in the world. We acknowledge the Bermuda Government Department of Conservation Services, the Bermuda Zoological Society and the Caribbean Conservation Corporation with Drs. Anne and Peter Meylan as scientific directors for the project. The Bermuda Amphibian Project, initiated in 1995, is a long-term investigation of the status and health of Bermuda's amphibian populations, the potential causes for their decline, and the implications for environmental health. The authors wish to thank the Bermuda Zoological Society (BZS) which hosts this project, The Bermuda Government Ministry of the Environment for its continued support, and the numerous charities, schools and businesses that have supported this project. Research was conducted in compliance with the Animal Welfare Act and other Federal statues and regulations relating to animals and experiments involving animals and adheres to principles stated in the Guide for Care and Use of Laboratory Animals (NRC, 1996). All of the recent research done on the Bermuda skink (Kitson in prep.) has been part of a Ph.D. program undertaken in cooperation with University College Cork, Ireland, under the supervision of Professor John Davenport. None of the research would have been possible had it not been for Professor Davenport and the generous assistance of the Bermuda Biodiversity Project, the Bermuda Zoological Society, the Bermuda Aquarium Museum and Zoo and Friends of the Bermuda Aquarium. L.K. would also like to thank Professor Tom Cross and Dr. Jamie Coughlan at University College Cork for their assistance with the genetics part of the research. This is Contribution No. 110 of the Bermuda Biodiversity Project (BBP), BAMZ.

References

Anderson, C., De Silva, H., Furbert, J., Glasspool, A., Rodrigues, L., Sterrer, W., Ward, J. (2001): Bermuda Biodiversity Country Study. Bermuda Zoological Society and The Bermuda Aquarium Museum and Zoo.

Babcock, H.L. (1937): Sea Turtles of the Bermuda Islands, with a Survey of the present state of the Turtle Fishing Industry. Proc. Zool. Soc. Lond. (A) **107**: 595-601.

Bacon, J.P., Linzey, D.W., Rogers, R.L., Fort, D.J. (2006): Deformities in cane toad (*Bufo marinus*) populations in Bermuda: Part I. Frequencies and distribution of abnormalities. Appl. Herpetol. **3**: 39-65.

Cadi, A., Joly, P. (2003): Competition for basking places between the endangered European pond turtle (*Emys orbicularis galloitalica*) and the introduced red-eared slider (*Trachemys scripta elegans*). Can. J. Zool. **81**: 1392-1398.

Cadi, A., Joly, P. (2004): Impact of the introduction of the red-eared slider (*Trachemys scripta elegans*) on survival rates of the European pond turtle (*Emys orbicularis*). Biodivers. Conserv. **13**: 2511-2518.

Carr, A., Carr, M., Meylan, A.B. (1978): The ecology and migrations of sea turtles, 7. The west Caribbean green turtle colony. Bull. Am. Mus. Nat. Hist. **162**: 1-46.

Coughlan, J.P., Kitson, L., Dillane, E., Davenport, J., Cross, T.F. (2004): Characterization of six microsatellite loci in the Bermuda skink (*Eumeces longirostris*). Mol. Ecol. Notes **4**: 678.

Conyers, J., Wingate, D. (1996): *Eumeces longirostris*. In: IUCN 2006. 2006 IUCN Red List of Threatened Species.

Davenport, J., Glasspool, A., Kitson, L. (2005): Occurrence of Diamondback Terrapins, *Malaclemys terrapin*, on Bermuda: Native or Introduced? Chelonian Cons. Biol. **4**: 956-959.

Davenport, J., Hills, J., Glasspool, A., Ward, J. (2001): Threats to the Critically Endangered endemic Bermudian skink *Eumeces longirostris*. Oryx **35**: 332-339.

Department of Agriculture and Fisheries Monthly Bulletin (1972): Sixth Year of the Green Turtle Restocking Project. September **42** (9).

Edgar, P., Kitson, L., Glasspool, A. (2006): Recovery Plan for the Bermuda Skink (*Eumeces longirostris*). Department of Conservation Services, Ministry of the Environment, Bermuda.

Engstrom, T.N., Bradley, W.G., Gray, J.A., Meylan, A.B., Meylan, P.A., Roess, W.B. (1998): Genetic identity of green turtles in Bermuda waters. In: Proceedings of the Seventeenth Annual Sea Turtle Symposium, p. 50. Epperly, S.P., Braun, J. (Compilers), U.S. Department of Commerce. NOAA Tech Memo. NMFS-SEFSC-415.

Fort, D.J., Thomas, J.H., Rogers, R.L., Bacon, J.P. (2006a): Deformities in cane toad (*Bufo marinus*) populations in Bermuda: Part II. Progress towards characterization of chemical stressors. Appl. Herpetol. **3**: 143-172.

Fort, D.J., Thomas, J.H., Rogers, R.L., Bacon, J.P. (2006b): Deformities in cane toad (*Bufo marinus*) populations in Bermuda: Part III. Microcosm-based exposure pathway assessment. Appl. Herpetol. **3**: 257-277.

Garman, S. (1884): The reptiles of Bermuda. Bull. U.S. Natl. Mus. **25**: 285-303.

Glasspool, A.F., Outerbridge, M. (2005): A Population Re-survey of the Bermuda Skink, *Eumeces longirostris* Cope (1861), on Southampton Island, Castle Harbour. BBP Special Publication 2005/001.

Gray, J.A., Meylan, A.B., Outerbridge, B. (2006): Two decades of stranding data from Bermuda, an island in the Sargasso Sea. In: 2006 Book of Abstracts, Twenty Sixth Annual Symposium on Sea Turtle Biology and Conservation, p. 296. Frick, M., Panagopoulou A., Rees, A.F., Williams, K. (Compilers). International Sea Turtle Society, Athens, Greece.

Gray J.A. (1990): Turtles Hatch at Clearwater. Crittertalk: Newsletter of the Bermuda Aquarium Museum and Zoo **13**: 3.

Gray, J.A., Mitchell, W.H., Ward, J.A., Frick II, H.C., Meylan, P.A., Meylan, A.B. (1998). The Bermuda turtle project: studies of immature green turtles at an oceanic feeding ground, 1968-1997. In: Proceedings of the Seventeenth Annual Sea Turtle Symposium, p. 59. Epperly, S.P., Braun, J. (Compilers). NOAA Technical Memorandum NMFS-SEFSC-415.

Griffith, H., Wingate, D.B. (1994): *Eumeces longirostris* (Bermuda rock lizard or skink). Predation. Herpetol. Rev. **25** (1): 26.

Heilprin, A. (1889): The Bermuda Islands: A Contribution to the Physical History and Zoology of the Somers Archipelago. Heilprin, Philadelphia.

Jones, J.M. (1859): The naturalist in Bermuda. Proc. Acad. Nat. Sci. Philadelphia **13**: 12-314.

Linzey, D.W., Burroughs, J., Hudson, L., Marini, M., Robertson, J., Nagarkatti, M., Bacon, J.P., Nagarkatti, P.S. (2003): Role of environmental pollutants on immune functions, parasitic infections and limb malformations in marine toads and whistling frogs from Bermuda. Int. J. Environ. Health Res. **13**: 125-148.

Linzey, D.W., Bursey, C.R., Linzey, J.B. (1998a): Seasonal occurrence of helminths of the whistling frog, *Eleutherodactylus johnstonei* (Amphibia: Leptodactylidae) in Bermuda. J. Helminthol. Soc. Wash. **65**: 245-251.

Linzey, D.W., Bursey, C.R., Linzey, J.B. (1998b): Seasonal occurrence of helminths of the giant toad, *Bufo marinus* (Amphibia: *Bufonidae*), in Bermuda. J. Helminthol. Soc. Wash. **65**: 251-258.

Meylan, A., Meylan, P., Mosier, A. (1994): Green turtles in developmental habitat: an update on the Bermuda Turtle Project. In: Proceedings of the 13th Annual Symposium on Sea Turtle Biology and Conservation, p. 254. Schroeder, B., Witherington, B. (Compilers). NOAA Technical Memorandum NMFS-SEFSC-341.

Meylan, P.A., Sterrer, W. (2000): *Hesperotestudo* (Testudines: Testudinidae) from the Pleistocene of Bermuda, with comments on the phylogenetic position of the genus. Zool. J. Linn. Soc. **128**: 51-76.

Meylan, P., Meylan, A. (1998): Corroboration of the developmental habitat hypothesis for marine turtles. In: Proceedings of the 17th Annual Symposium on Sea Turtle Biology and Conservation, p. 68. Epperly, S.P., Braun, J. (Compilers). NOAA Technical Memorandum NMFS-SEFSC-415.

Meylan, P., Meylan, A., Gray, J. (1992, 1994, 1999, revised 2003): BAMZ # 1139 Procedures Manual for the Bermuda Turtle Project.

Meylan, P., Meylan, A., Gray, J., Ward, J. (2004): The hawksbill turtle in Bermuda (abstract). Proceedings of the 22nd Annual Symposium on Sea Turtle Biology and Conservation. NOAA Technical Memorandum NMFS-SEFC-503:26.

Mowbray, L.S., Caldwell, D.K. (1958): First record of the ridley turtle from Bermuda, with notes on other sea turtles and the turtle fishery in the islands. Copeia **1958**: 147-148.

Pope, P.H. (1917): The introduction of West Indian *Anura* into Bermuda. Bull. Comp. Zool. **61**: 117-131.

Raine, A. (1998): A study of the morphological differentiation, fluctuating asymmetry and the threats facing isolated populations of the critically endangered Bermuda Rock Lizard (*Eumeces longirostris*). M.Sc. dissertation, University College, London.

Royal Gazette (1993): Island's toads are frogs dying out says scientist. May 10.

Samuel, D. (1975): Feeding habits of the kiskadee flycatcher. Bermuda Biol. Stn Newsl. **4**: 3-4.

Thomas, M.L.H. (2004): The Natural History of Bermuda. Brimstone Media Ltd, Bermuda.

Verrill, A.E. (1902): The Bermuda Islands: An account of their scenery, climate, productions, physiography, natural history and geology, with sketches of their discovery and early history, and the changes in the flora and fauna due to man. 2nd edition. Trans. Connecticut Acad. Sci. **11**.

Verrill, A.E. (1905): The Bermuda Islands Pt. IV. Geology and Paleontology. Trans. Connecticut Acad. Sci. **12**.

Vierros, M., Meylan, A., Meylan, P., Gray, J., Ward, J. (2002): Evaluation of green turtle habitat, population size and distribution using remote sensing and GIS Techniques. In: Proceedings of the Twentieth Annual Sea Turtle Symposium, p. 52-54. Mosier, A.F., Brost, B. (Compilers). NOAA Technical Memorandum NMFS-SEFSC-477.

Wingate, D.B. (1965): Terrestrial herpetofauna of Bermuda. Herpetologica **21**: 202-218.

Wingate, R. (1998): A comparison of demography and morphological variation in two insular populations of the Bermuda rock lizard (*Eumeces longirostris*). B.Sc. dissertation. University of Swansea, Wales.

Accepted: August 22, 2006 (AH).
Reprinted from Applied Herpetology **3**: 323-344 (2006).

Conservation of herpetofauna in the Republic of Trinidad and Tobago

Adrian Hailey[1,2], Michelle Cazabon-Mannette[1,3]

[1] Department of Life Sciences, The University of the West Indies, St. Augustine, Trinidad and Tobago, West Indies
[2] Corresponding author; e-mail: adrian.hailey@sta.uwi.edu
[3] Turtle Village Trust, PO Box 1109, Wrightson Rd, Port of Spain, Trinidad and Tobago, West Indies

Abstract. Trinidad and Tobago present three main contrasts to other Caribbean islands in terms of herpetological conservation. (1) They are continental shelf rather than oceanic islands and have a diverse Neotropical fauna and a low level of endemism. (2) They were developed relatively late in the European colonial period, and the delayed population and agricultural growth has left relatively large areas of original habitat. (3) Industrial development is the major feature of current economic growth in the Republic of Trinidad and Tobago. Trinidad is more divergent from the typical Caribbean island than Tobago in all these respects. The herpetofauna is incompletely known, but about 112 terrestrial and freshwater species and five sea turtles have been recorded. Trinidad has 103 species, Tobago 52, with 43 species in common; 83% of the species are also found in South America. There are 12 endemic and seven introduced species. Iguana and tegu lizards, as well as caiman, are hunted under licence, and there is a legal artisanal fishery for sea turtles. Venomous snakes (absent from Tobago) are classed as vermin and may legally be destroyed. Other reptiles are classed as protected animals under the Conservation of Wild Life Act; amphibians are currently unprotected by law outside of Environmentally Sensitive Areas. The amphibian chytrid fungus is present on both islands. The endemic Critically Endangered (IUCN) golden tree frog *Phytotriades auratus* will probably receive protection as an Environmentally Sensitive Species in the near future; this status was recently rejected for the sea turtles. There is effective practical conservation of nesting turtles by community-based ecotourism projects on some beaches, recently coordinated by the Turtle Village Trust, and efforts to replace the turtle fishery by non-destructive economic use as an attraction for dive ecotourism. A large bycatch of leatherback turtles in the gillnet fishery remains a major conservation issue. Future problems related to the development of industry and effects of global climate change are discussed, and recommendations given for conservation of the herpetofauna.

Key words: *Batrachochytrium dendrobatidis*; conservation legislation; *Dermochelys coriacea*; ecotourism; hunting; incidental catch; Neotropics; *Phytotriades auratus*; protected areas; sea turtle fishery.

Introduction

Trinidad is an island of 4830 km^2, separated from Venezuela by the Gulf of Paria which is 11 km wide at the narrowest point; Tobago has an area of 300 km^2 and lies a further 36 km to the north-east (fig. 1). The shallowest channel between Trinidad and Tobago has a maximum depth of 91 m, and that between Trinidad and Venezuela of 38 m, so both islands would have been connected to the mainland during the last glacial maximum when sea level was 135 m lower (Clark and Mix, 2002). Based on channel depths and sea levels, the last separation was 14,000 BP for Tobago and 10,000 BP for Trinidad. An alternative view, based on the presence of coral remains in the Gulf of Paria, is that Trinidad was connected to Venezuela until only 1500 years ago by a land bridge, until that was breached by the Orinoco to form the south-west peninsula (Kenny, 1989, 1995). In either case, the date of separation is too recent to account for speciation, and molecular evidence indicates that endemic species in Trinidad and Tobago formed 1-5 MYA (e.g., Manzanilla et al., 2009). Murphy (1997) estimated that Trinidad was connected to the mainland for 65% of the last 140,000 years, and Tobago for 14%. The Northern Range of Trinidad (fig. 2a) is composed of Mesozoic metamorphic rocks, representing the end of the Andes mountain chain. It has a maximum height of 940 m at Mt Aripo and 936 m at El Tucuche. The Central Range of younger sedimentary rocks has a maximum height of 307 m at Mount Tamana; the Southern Range or Trinity Hills is of even

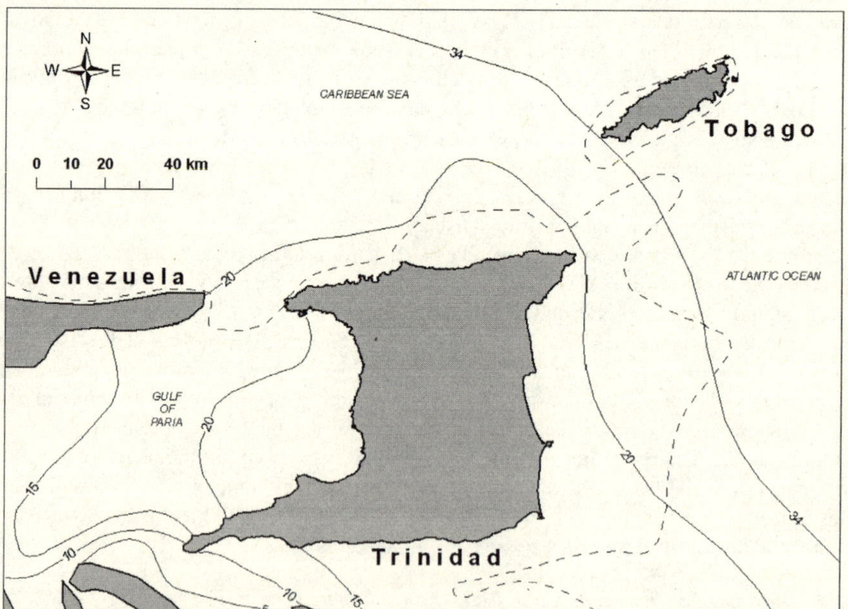

Figure 1. Trinidad, Tobago, and the adjacent coast of Venezuela. Solid lines show wet season surface salinity levels (ppt) and the dashed line shows the 54 m (30 fathom) depth contour (after Kenny, 1995, and Spalding, 2004).

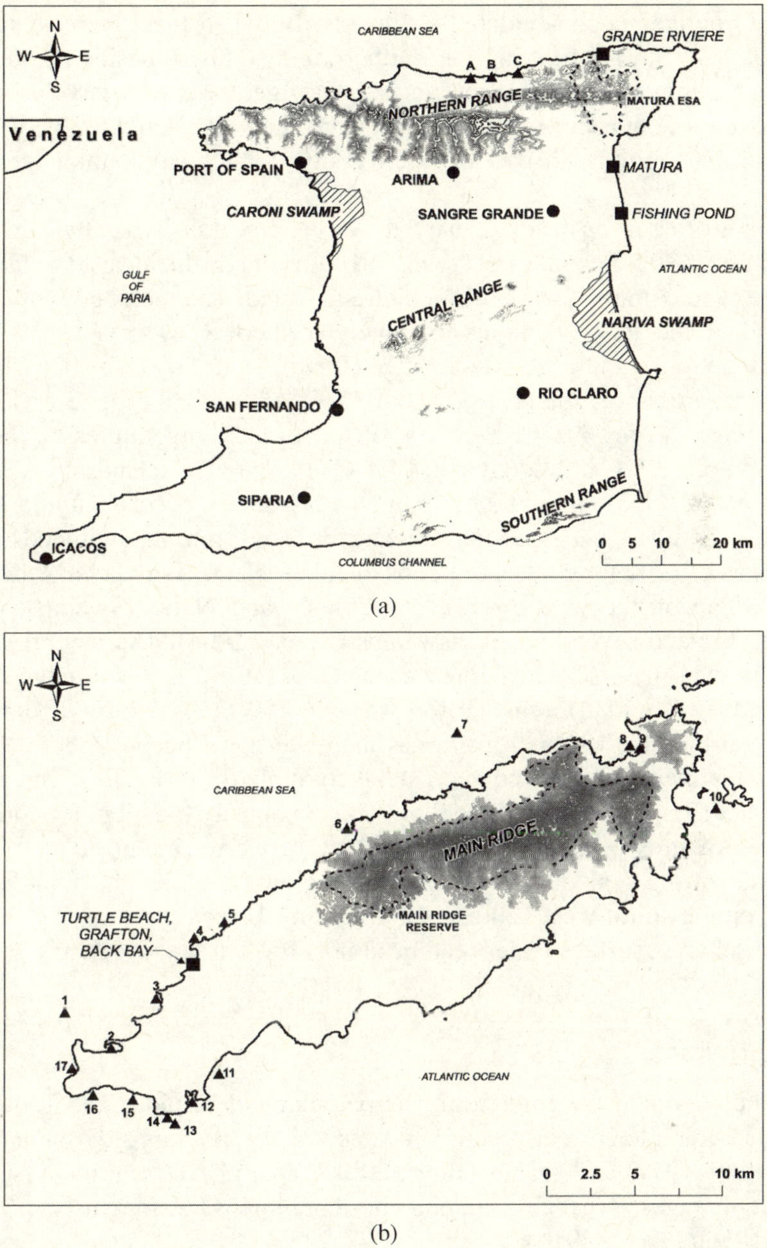

Figure 2. (a) Trinidad, with highlands above 100 m and the major swamps. The dashed line shows the Matura Environmentally Sensitive Area. A, Paria; B, Tacarib; C, Madamas. The six Wildlife Section offices issuing the State Game Licence are at Port of Spain, Arima, Sangre Grande, Rio Claro, San Fernando and Siparia. (b) Tobago, with highlands above 300 m. The dashed line shows the Main Ridge Reserve. 1, Buccoo Reef; 2, Bon Accord Lagoon; 3, Mt. Irvine; 4, Plymouth; 5, Arnos Vale; 6, Castara; 7, Sisters Rocks; 8, Charlotteville; 9, Man-O-War Bay; 10, Speyside and Little Tobago; 11, Canoe Bay; 12, Petit Trou; 13, Cove; 14, Little Rockley Bay; 15, Kilgwyn; 16, Crown Point; 17, Store Bay.

lower and younger unconsolidated sediments. Between these ranges are alluvial terraces of 15-60 m asl forming the northern and southern basins (Liddle, 1946). Tobago has a more complex (and controversial) geology; it is largely of igneous island-arc origin, but accreted to the continental shelf (Frost and Snoke, 1989). The highlands in Tobago (fig. 2b) are the Main Ridge, with a maximum height of about 550 m.

There is a dry season from January to May; some rain typically falls in every month, but over 80% is in the wet season. In both islands the lowest rainfall is in the west, and the maximum occurs on the highest ground. The recorded spatial range is 1.2-3.8 m in Trinidad, probably up to 5 m on Northern Range peaks, and 1.4-2.4 m in Tobago, up to 3.8 m on the Main Ridge (Beard, 1946; MPUE, 2006). The mean ambient temperature at sea level is about 27°C, with a daily range of 11°C but a seasonal range of only 2-3°C. Beard (1946) recorded temperatures on the summit of Mt Aripo of 17.8°C at midday and 14.4°C at night. The islands are south of the hurricane belt and experience them rarely; the last severe events being in Tobago in 1963 and in Trinidad in 1933. Many streams descend from the highlands, but there are no large rivers, natural lakes or ponds in either island. Trinidad has the brackish Caroni Swamp on the west coast and the freshwater Nariva Swamp on the east coast (fig. 2a), and several smaller swamps (Kenny, 2008). The natural vegetation is typically evergreen seasonal forest up to 200-250 m, lower montane forest to 750 m, and (in Trinidad) montane forest above 750 m and elfin woodland above 850 m (Beard, 1944, 1946). Tobago has the extensive Buccoo Reef in the south-west, with that part of the island being a flat coral platform (fig. 2b). The salinity of the sea around Trinidad is mostly too low, and the opacity too high, for coral growth due to the proximity of the Orinoco mouth, but there are specialized reefs along the north coast. During peak discharge, salinity is below 20 ppt around Trinidad, falling to 15 ppt on the south-west peninsula and to only 10 ppt at its tip (fig. 1). During extreme discharge, surface waters can be almost fresh at Icacos (Kenny, 2008).

The Herpetofauna

There is a comprehensive guide to the herpetofauna of Trinidad and Tobago (Murphy, 1997), which is currently being updated, and monographs on the amphibians (Kenny, 1969, 1977) and snakes (Boos, 2001). There have been many subsequent taxonomic changes, especially among the amphibians. Names used here follow Frost (2010) and Uetz (2010).

The terrestrial and freshwater herpetofauna comprises 112 species (subject to interpretation and new discoveries): 35 anurans, 1 crocodilian, 6 chelonians, 2 amphisbaenians, 26 lizards, and 42 snakes. Trinidad has 103 species and Tobago has 52, with 43 species in common. Tobago has no venomous snakes — hence the normally mimetic genus *Erythrolamprus* is represented by *E. ocellatus*, a form with ocelli rather than bands (Emsley, 1966). Jaccard's binary index of similarity $S_j = a/(a + b + c)$ (where a is the number of species common to both islands, and *b* and *c* are the number of species found on only one island; Krebs, 1999) is equal to 0.38 overall and 0.39 excluding introduced species. Sorensen's index ($S_s = 2a/(2a + b + c)$) is 0.55 overall, and 0.56 excluding introduced species. The latter index weights similarities more than differences, and is equal to Duellman's Faunal Resemblance Factor used by Murphy (1997). The similarity between the herpetofaunas of Trinidad and of Tobago is thus limited, largely due to the absence of some species from Tobago, rather than a high proportion of island endemics.

The great majority (93, or 83%) of the 112 species are also found in South America (table 1). Some other Neotropical species may also be, or have been at one time, part of the Trinidad herpetofauna. *Leptodactylus knudseni* was recorded from a single specimen before 1919, but may be present as subfossil material associated with humans. This large frog may have been extirpated by humans or by the mongoose. *Crocodylus acutus* was also possibly extirpated by humans, and there are records of a *Typhlonectes* sp. (caecilian), the river turtles *Podocnemis expansa* and *P. unifilis*, *Anolis* cf. *lemurinus*, the gecko *Gymnodactylus geckoides* and the colubrid *Erythrolamprus bizona* in Trinidad (Murphy, 1997). On the other hand, Kenny (2008) suggests that some species accepted as part of the Trinidad herpetofauna may not form breeding populations, but be vagrant individuals from the Orinoco; e.g., *Chelus fimbriatus*, which are often covered in barnacles indicating passage in brackish water.

Twelve species are endemic to Trinidad and/or Tobago (table 2a). Recent work on *Mannophryne trinitatis* shows that the Venezuelan populations are distinct from those in Trinidad (Manzanilla et al., 2007), which thus become an endemic species. As with most of the Caribbean herpetofauna (Wilson et al., 2006) the endemic species of Trinidad and Tobago are little known; only *M. trinitatis* can be considered well-studied, especially its tadpole transport and deposition behaviour (e.g., Downie et al., 2005; Jowers and Downie, 2005). This species is widely distributed in the Northern and Central Ranges (Jowers and Downie, 2004) and shows variation in ecology (Cummins and Swan, 1995) and behaviour (Jowers et al., 2006) between populations. The similar Bloody Bay frog *M. olmonae* is a Tobago endemic. *Pristimantis urichi* and *Typhlops trinitatus* are widespread in both islands. The golden tree frog *Phytotriades auratus* has recently been assigned to a new genus separate from *Phyllodytes* (Jowers et al., 2008); such generic recognition will help to promote *P. auratus* as an important endemic worthy of conservation. It is only found at high elevation, on Mt Aripo, El Tucuche, and (perhaps formerly) on other high peaks (Kenny, 2008). Its lower elevation limit is between 800-900 m asl, apparently corresponding to that of its breeding habitat, the giant tank bromeliad *Glomeropitcairnia erectiflora*, although individuals have occasionally been found in other bromeliad species (and will use them, and artificial bromeliads, in captivity; A. Hailey, pers. obs.). The so-called "luminous lizard" (Knight et al., 2004), *Riama shrevei*, also has a montane distribution in the Northern Range, down to about 600 m asl, and is mostly associated with caves at the lower levels. *Leptodactylus nesiotus* is known from a few specimens from the south-west peninsula of Trinidad, and

Table 1. Native terrestrial and freshwater herpetofauna species in Trinidad and Tobago having a South American distribution. The number of species in each area, and the IUCN Red List category or CITES Appendix (CA), are shown in parentheses. Red List categories are: EN = Endangered; VU = Vulnerable; LC = Least Concern; LR/lc = Lower Risk/least concern; LR/cd = Lower Risk/conservation dependent. *Species, genus and/or family name has changed since Murphy (1997).

Both Trinidad and Tobago (39)

Bufonidae: *Rhinella marina** (LC).

Amphignathodontidae: *Flectonotus fitzgeraldi** (EN).

Hylidae: *Dendropsophus minutus** (LC); *Hypsiboas crepitans** (LC); *Scinax ruber** (LC); *Trachycephalus venulosus** (LC).

Leiuperidae: *Engystomops pustulosus** (LC).

Leptodactylidae: *Leptodactylus fuscus* (LC); *Leptodactylus validus* (LC).

Alligatoridae: *Caiman crocodilus* (LR/lc, CAII).

Gekkonidae: *Gonatodes vittatus*; *Hemidactylus palaichthus*; *Sphaerodactylus molei*; *Thecadactylus rapicauda*.

Gymnophthalmidae: *Bachia heteropa*.

Iguanidae: *Iguana iguana* (CAII).

Polychrotidae: *Polychrus marmoratus**.

Scincidae: *Mabuya bistriata*.

Teiidae: *Ameiva ameiva*; *Cnemidophorus lemniscatus*; *Tupinambis teguixin* (CAII).

Boidae: *Boa constrictor* (CAII); *Corallus hortulanus* (CAII); *Epicrates cenchria* (CAII).

Colubridae: *Atractus trilineatus*; *Drymarchon corais*; *Imantodes cenchoa*; *Leptodeira annulata*; *Leptophis ahaetulla*; *Liophis melanotus*; *Liophis reginae*; *Mastigodryas boddaerti*; *Ninia atrata*; *Oxybelis aeneus*; *Oxyrhopus petola*; *Pseudoboa neuwiedii*; *Sibon nebulata*; *Spilotes pullatus*; *Tantilla melanocephala*.

Trinidad Only (51)

Bufonidae: *Rhinella beebei** (LC).

Hylidae: *Dendropsophus microcephalus** (LC); *Dendropsophus minisculus** (LC); *Hypsiboas boans** (LC); *Hypsiboas geographicus** (LC); *Hypsiboas punctatus** (LC); *Phyllomedusa trinitatis* (LC); *Pseudis paradoxa** (LC); *Sphaenorhynchus lacteus* (LC).

Leptodactylidae: *Leptodactylus bolivianus* (LC); *Leptodactylus hylaedactylus** (LC); *Leptodactylus lineatus** (LC); *Leptodactylus ocellatus** (= *L. macrosternum*) (LC).

Microhylidae: *Elachistocleis ovalis* (LC); *Elachistocleis surinamensis* (LC).

Pipidae: *Pipa pipa* (LC).

Ranidae: *Lithobates palmipes** (LC).

Chelidae: *Chelus fimbriatus*; *Mesoclemmys gibba**.

Bataguridae: *Rhinoclemmys punctularia**.

Kinosternidae: *Kinosternon scorpioides*.

Testudinidae: *Chelonoidis denticulata** (VU, CAII).

Amphisbaenidae: *Amphisbaena alba*; *Amphisbaena fuliginosa*.

Gekkonidae: *Gonatodes ceciliae*; *Gonatodes humeralis*.

Gymnophthalmidae: *Gymnophthalmus speciosus*; *Gymnophthalmus underwoodi*.

Polychrotidae: *Anolis chrysolepis**.

Teiidae: *Kentropyx striata**.

Tropiduridae: *Plica plica**.

Boidae: *Eunectes murinus*.

Colubridae: *Chironius carinatus*; *Clelia clelia* (CAII); *Dipsas variegata*; *Erythrolamprus aesculapii*; *Helicops angulatus*; *Hydrops triangularis*; *Leptophis riveti*; *Liophis cobella*; *Pseustes poecilonotus* (LC); *Pseustes sulphureus*; *Siphlophis cervinus*; *Siphlophis compressus**; *Thamnodynastes* sp.

Elapidae: *Micrurus circinalis*; *Micrurus lemniscatus*.

Table 1. (Continued).

Leptotyphlopidae: *Epictia tenella** (= *Leptotyphlops albifrons*).
Typhlopidae: *Typhlops brongersmianus*.
Viperidae: *Bothrops atrox**; *Lachesis muta*.

Tobago Only (3)
Centrolenidae: *Hyalinobatrachium orientale* (VU).
Gymnophthalmidae: *Bachia flavescens*.
Colubridae: *Atractus univittatus*.

the anomalepid blind snake *Helminthophis* sp. from a single specimen. The four endemic amphibians in Tobago all have similar recorded distributions, in the north-east of the island, but are possibly more widespread. In addition to endemic species, there are probably endemic subspecies of *Bachia heteropa*, *Liophis reginae* and *Mastigodryas boddaerti* on Tobago, and of *Dipsas variegata* on Trinidad (Murphy, 1997).

The only Antillean elements in the Trinidad and Tobago herpetofauna are the human introductions, *Eleutherodactylus johnstonei* and *Anolis* spp. Seven introduced species of herpetofauna are currently known from Trinidad and/or Tobago (table 2b), and another species (*Anolis extremus*) is now extinct in Trinidad (Hailey et al., 2009). *Hemidactylus mabouia* is nearly cosmopolitan, but apparently originated from Africa (Carranza and Arnold, 2006); it may have displaced the native *H. palaichthus* which is currently known only from small offshore islands (Monos and Chacachacare, off north-west Trinidad, and from Little Tobago). *Chelonoidis carbonaria* is likely to have been an ancient introduction for food. The most recent introductions *E. johnstonei* and *Anolis wattsi* are spreading rapidly in suburban areas (Manickchan, 2003; White and Hailey, 2006).

The herpetofauna of Trinidad and Tobago is incompletely documented. There are possibly two (at least) additional frogs; a new *Pristimantis* species in the Northern Range (M. Patrikeev and S.B. Hedges, pers. comm., April 2010) and a Scarthylid in the south-west peninsula (Ogilvy et al., 2007; Smith, 2007). *Anolis aeneus* has recently been found in Tobago (G. White, pers. comm., 2 April 2008). Natural invasions of Orinoco species are always possible, from the strong flow of almost fresh water reaching the south-west peninsula at times; this may be responsible for the several species found only (as yet) in that area. Anacondas are frequently washed up (e.g., Charan, 2007), and rafting amphibians are a strong possibility.

Sea turtles

Five species of sea turtles occur in Trinidad and Tobago.

Dermochelys coriacea. Gravid female leatherbacks are seasonal visitors to the Caribbean (males are rarely encountered) and observations are largely confined to the peak breeding months from March to August (Eckert, 2001). There is a large breeding population using beaches in Trinidad, mostly on the north and east coasts

Table 2. Endemic and introduced species of terrestrial and freshwater herpetofauna in Trinidad and Tobago. The number of species in each area and the IUCN Red List category or CITES Appendix (CA), are shown in parentheses. Red List categories are: CR = Critically Endangered; EN = Endangered; VU = Vulnerable; LC = Least Concern. *Species, genus and/or family name has changed since Murphy (1997).

(a) Endemic Species

Both Trinidad and Tobago (2)	Trinidad Only (5)	Tobago Only (5)
Strabomantidae:	Aromobatidae:	Aromobatidae:
*Pristimantis urichi** (EN)	*Mannophryne trinitatis** (VU)	*Mannophryne olmonae** (CR)
Typhlopidae:	Hylidae:	Strabomantidae:
Typhlops trinitatus	*Phytotriades auratus* (CR)	*Pristimantis charlottevillensis** (VU)
		*Pristimantis turpinorum** (VU)
	Leptodactylidae:	
	Leptodactylus nesiotus (VU)	
	Gymnophthalmidae:	Gekkonidae:
	*Riama shrevei**	*Gonatodes ocellatus*
	Anomalepidae:	Colubridae:
	Helminthophis sp.	*Erythrolamprus ocellatus*

(b) Introduced Species

Both Trinidad and Tobago (2)	Trinidad Only (4)	Tobago Only (1)
Gekkonidae:	Eleutherodactylidae:	Polychrotidae:
Hemidactylus mabouia	*Eleutherodactylus johnstonei** (LC)	*Anolis richardii**
Polychrotidae:	Testudinidae:	
*Anolis aeneus**	*Chelonoidis carbonaria** (CAII)	
	Polychrotidae:	
	*Anolis trinitatis**	
	*Anolis wattsi**	

(Pritchard, 1984). The current figure is about 6000 females annually (Eckert, 2006), making the nesting population the largest in the insular Caribbean and perhaps the second largest in the World (Fournillier and Eckert, 2009). A total of 5642 nesting events by 3757 females were recorded at monitored beaches in 2009; 2176 at Grande Riviere, 2212 at Matura, 761 at Fishing Pond, and the remainder in Tobago (Turtle Village Trust, unpubl. data). Large numbers of leatherbacks also nest on the more remote beaches west of Grande Riviere, especially Paria, Madamas, and Tacarib (fig. 2a; Livingstone, 2006). The leatherback is said to be unpalatable and of little commercial value, although the forelimb muscles may be eaten (Fisheries Division Data Collector at Guayaguayare, pers. comm., 12 March 2004). There was widespread slaughter of leatherbacks on north and east coast beaches in the 1960s (Bacon, 1970), apparently due to vandalism; most of the carcasses were left or only used as shark fishing bait (Gaskin and Shephard, 1994; Lambie, 2005). The species has since been effectively protected in Trinidad on the three main nesting beaches, but is subject to poaching and harassment on unprotected beaches. There is a small nesting population on the west coast of Tobago, subjected to persistent poaching; Bacon (1981) and Nathai-Gyan et al. (1987) estimated the population to number in the dozens. Since then the population has increased, the most important nesting beaches being Turtle Beach and Grafton Beach (Fournillier and Eckert, 1998). Substantial slaughter of leatherbacks at Black Rock in 1999 led to the formation of Save Our Sea Turtles Tobago, which has monitored turtle beaches since 2000 (Clovis, 2004); 378 leatherback nests were recorded in 2008, and 421 in 2009 (Lalsingh, 2008, 2009). Egg poaching (of any sea turtle) is known in both islands but is not as common as slaughter of females on the beach and is not considered a major problem.

Caretta caretta. The major breeding areas of the loggerhead in the Atlantic are on the south-east coast of the USA. Loggerheads undertake transoceanic journeys as juveniles, travelling in the Gulf Stream to the eastern Atlantic and back in the Northern Equatorial Current, passing north of Trinidad (Gavilan, 2001). Local data are consistent with this pattern; loggerheads rarely nest in Trinidad (Bacon, 1981), and are only occasionally caught by fishermen off the north coast (Gaskin and Shephard, 1994). There is no evidence of nesting or foraging in Tobago (Fournillier and Eckert, 1998).

Chelonia mydas. The green turtle is a resident foraging species in Trinidad, also occasionally nesting on the north-east, west and south coasts (Pritchard, 1984), and is a principal target of the turtle fishery (in which it is known as the greenback). In Tobago the seagrass meadow foraging habitat is concentrated around the south-west (Fournillier and Eckert, 1998). Sightings of green turtles have been made all around the island (fig. 2b, sites 1-3, 5, 9, 10, 12-15, 17; M. Cazabon-Mannette, unpubl. data). Six nesting green turtles were observed in Tobago from 2000-2004 (Clovis, 2004).

Eretmochelys imbricata. The hawksbill nests on the north and east coasts of Trinidad (Pritchard, 1984) and in Tobago (Lalsingh, 2008, 2009). Livingstone (2006) reported about 675 nests of hawksbills on the north coast of Trinidad each year from 2000-2004. The extent of hawksbill foraging habitat is probably greater in Tobago than in Trinidad. Cazabon-Mannette (unpubl. data) found hawksbills to be distributed on reefs between Cove and Crown Pt. on the south-west of the island (fig. 2b, sites 13-16), along the north coast (sites 3-8), and throughout Speyside. The hawksbill was formerly utilized by the turtle fishery in Trinidad and Tobago for the carapace, used as a source of tortoiseshell, as well as for meat. Chu Cheong (1995) reported the value of carapace to be several times that of meat. The export of hawksbill shell is now illegal (Gaskin and Shephard, 1994).

Lepidochelys olivacea. The olive ridley (or batali in Trinidad) is probably the most abundant sea turtle worldwide, found in all tropical and subtropical ocean basins, but is the least abundant species in the western Atlantic (Marcovaldi, 2001). It is rare in Trinidad (Bacon, 1981), but possibly nests on the west (Pritchard, 1984; Godley et al., 1993) and north (Livingstone, 2005) coasts, and has been reported in offshore nets. Sightings are very infrequent in Tobago; confusion in reporting is possible because batali refers to the leatherback in Tobago (Fournillier and Eckert, 1998).

Conservation status

Nine species of the terrestrial and freshwater herpetofauna are listed by CITES (tables 1 and 2), all of them on Appendix II (CITES, 2010). *Caiman*, *Iguana* and *Tupinambis* are hunted locally, but there is no export trade — movement of wild meat is generally in the opposite direction with imports (often illegal) from South America, though reptiles have not featured in this trade to date. Apart from the colubrid *Clelia*, the other species on CITES Appendix II are all part of blanket international restrictions on trade in boas and tortoises. *Podocnemis* species, that are probably at least a transient part of the Trinidad herpetofauna, are also on Appendix II. All sea turtles are listed on Appendix I of CITES, with commercial international trade prohibited.

Table 3 summarizes the IUCN Red List (IUCN, 2010) categories and criteria for the 15 species listed as Vulnerable, Endangered, or Critically Endangered; these are all amphibians or chelonians. The criteria are based on declining population size (A), limited geographic range (B), or small population size (D), with a clear distinction between the two taxa in the causes of endangerment.

The threatened amphibians generally owe their status to restricted geographic range (B1), due to a small number of locations (a) or declining area (b_{iii}), or because the small range makes populations vulnerable to disturbance or stochastic events (D2); one of these factors applies in all but one case. The exception is *P. urichi*, the threat status of which is solely due to a substantial reduction in population size (A2), and which is categorized as Endangered. Additional factors which lead to

Table 3. IUCN Red List categories[1] and criteria[2] for threatened herpetofauna in Trinidad and Tobago. *Species in which the genus has changed since Murphy (1997).

	Category	A1				A2					B1			B2		D2
		a	b	c	d	a	b	c	d	e	a	$b_{(iii)}$	$b_{(v)}$	a	$b_{(iii)}$	
Hyalinobatrachium orientale	VU 2001	–	–	–	–	–	–	–	–	–	•	•	–	–	–	–
Mannophryne olmonae	CR 2001	–	–	–	–	•	–	–	–	•	•	–	•	–	–	–
Mannophryne trinitatis	VU 2001	–	–	–	–	–	–	–	–	–	•	•	–	–	–	–
Flectonotus fitzgeraldi	EN 2001	–	–	–	–	–	–	–	–	–	•	•	–	–	–	–
*Phytotriades auratus**	CR 2001	–	–	–	–	–	–	–	–	–	•	•	–	•	•	–
*Pristimantis charlottevillensis**	VU 2001	–	–	–	–	–	–	–	–	–	–	–	–	–	–	•
*Pristimantis turpinorum**	VU 2001	–	–	–	–	–	–	–	–	–	–	–	–	–	–	•
*Pristimantis urichi**	EN 2001	–	–	–	–	•	–	–	–	•	–	–	–	–	–	–
Leptodactylus nesiotus	VU 2001	–	–	–	–	–	–	–	–	–	–	–	–	–	–	•
Dermochelys coriacea	CR 1994	•	•	–	•	–	–	–	–	–	–	–	–	–	–	–
Caretta caretta	EN 1994	•	•	–	•	–	–	–	–	–	–	–	–	–	–	–
Chelonia mydas	EN 2001	–	–	–	–	–	•	–	•	–	–	–	–	–	–	–
Eretmochelys imbricata	CR 2001	–	–	–	–	–	•	–	•	–	–	–	–	–	–	–
Lepidochelys olivacea	VU 2001	–	–	–	–	–	•	–	•	–	–	–	–	–	–	–
*Chelonoidis denticulata**	VU 1994	–	–	•	•	–	–	•	•	–	–	–	–	–	–	–

[1] Red List categories are: CR = Critically Endangered; EN = Endangered; VU = Vulnerable, based on 1994 (version 2.3) or 2001 (version 3.1) categories and criteria (IUCN, 2010).

[2] Criteria may be briefly summarized as follows, for the 2001 version unless specified; details of thresholds (population declines, areas, population sizes) differ among categories, being more severe for higher (CR) and less severe for lower (VU) categories. Criterion A is a decline in population size over 10 years or three generations, whichever is the longer. Criterion A1 specifies a more severe decline than A2, but the latter is exacerbated by being ongoing, irreversible, or not understood. Factors on which the decline is based are (a) direct observation of numbers, (b) an index of abundance, (c) the area of occupancy, (d) exploitation, (e) effects of pathogens, introduced taxa etc. In the 1994 version A1 is an observed past decline while A2 is a projected decline over the next 10 years or three generations. Criterion B is limited geographic range, in terms of total range (B1, extent of occurrence) or area of habitats utilized within that range (B2, area of occupancy). Factors involved are (a) fragmentation or small number of locations, (b) continuing decline of range, based on (b_{iii}) area or quality of habitat, or (b_v) number of mature individuals. Criterion D is small population size, with D2 for Vulnerable species (only) being a small area of occupancy or number of locations making them prone to disturbance or stochastic events.

higher threat categories are a small area of utilized habitat (B2) in the case of *P. auratus*, and a substantial population size reduction in *M. olmonae*, both of which are Critically Endangered. In both cases of declining population size, the factors involved in this assessment are (a) direct evidence of decline, and (e) effects of pathogens. Both species are described as having declined within the last 5-10 years in protected forest areas, possibly due to the amphibian chytrid fungus (IUCN, 2010). The decline of *P. urichi* is reported only for Tobago, not Trinidad. Evidence for these population declines is, however, anecdotal. The taxonomic separation of *M. trinitatis* from populations in Venezuela greatly reduces the geographic range of the species. Together with the finding of the amphibian chytrid fungus in *M. trinitatis*, these factors are likely to lead to change in its Red List category, perhaps to Endangered.

In contrast to the amphibians, the threatened chelonians owe their status to declining population size (A1 and/or A2), determined either by direct observation of numbers (a) or through an index of abundance (b) in sea turtles, and by decline in habitat (c) of *C. denticulata*. The most threatened are *D. coriacea* and *E. imbricata*, which are Critically Endangered. The reason for declining population size is listed as exploitation (d) in all cases, rather than pathogens as in the amphibians.

Interactions with Humans

Environmental change in Trinidad and Tobago

Trinidad was first settled about 8000 years ago, from the Orinoco delta (Wilson, 2007). It was discovered by Europeans in 1498, by Columbus, who supposedly also sighted Tobago. The islands had been inhabited by a sequence of coastal Amerindian groups, who hunted sea turtles, lizards, tortoises, snakes, and caiman, in addition to mammals, birds and fish (Boomert, 2000; Steadman and Stokes, 2002). Trinidad and Tobago were still covered in forest when discovered by Europeans.

Sea turtles accounted for the largest number of identified specimens at two Amerinidian sites in Tobago (Steadman and Stokes, 2002). It is probable that effects on other herpetofauna were low, although small species seem to have been used to a greater extent in Tobago as larger game declined (Steadman and Jones, 2006). Thus, reptiles made up 46% of identified specimens at the earlier (2900 BP) site; 94% of those were sea turtles, with the rest being *Iguana*, *Boa*, and unidentified colubrid(s). At a later (1200-900 years old) site, reptiles made up 45% of identified specimens, but sea turtles only accounted for 42% of those, with *Iguana* (31%), and *Boa* (15%) being more frequent, along with smaller numbers of *Tupinambis*, *Ameiva*, *Polychrus*, *Rhinella marina*, unidentified colubrid snakes and small lizards (Steadman and Stokes, 2002).

Trinidad was part of the Spanish Empire until 1797 when it was captured by Britain. Tobago had a complex history involving several European countries, but was British from 1815. Economic development (Williams, 1942), population growth, and probably effects on the herpetofauna were low at that time, perhaps with the exception of tortoises (morocoys). A 1595 Spanish source quoted by Boomert (2000) reported a "great store of tortoises" in Trinidad, some being received from Amerindians. The population of Trinidad was only 36,600 in 1838, while Jamaica (about twice the area) had a population ten times greater in 1844 (Watts, 1987). The early economy was based on sugar, with annual production of 4900 tons for Trinidad and 7000 tons for Tobago in 1800; Jamaica and the Lesser Antilles had both reached comparable production a full century earlier (Watts, 1987). Nevertheless, environmental damage was sufficient to spur protection of the Main Ridge of Tobago in 1776, the oldest legally protected forest reserve of its kind in the world. The aim was to "remove to Your Majesty a tract of Wood Land lying in the interior and most hilly parts of this island for the purpose of attracting frequent Showers of Rain upon which the Fertility of Lands in these Climates doth entirely depend" (as quoted by Environment Tobago, 2010).

Sugar production in Trinidad increased rapidly in the 19th Century, while it declined from the 1800 level in Tobago. Population size increased to 258,000 in 1901, which included immigration of indentured workers from India from 1845-1917. The 19th Century saw most of the suitable lowland areas being brought into cultivation, but because of the late start compared to other Caribbean islands, much of the highlands remained undisturbed (Watts, 1987). Trinidad and Tobago were joined administratively within the British Empire in 1889. Tobago previously had a local democratic assembly while Trinidad had Crown Colony status and was governed directly from London. A lasting consequence is that Tobago today has considerable local autonomy through the Tobago House of Assembly (THA), especially in environmental matters.

The 20th Century saw the continuing decline of sugar production in Tobago, the industry being closed in the 1930s; as a result Tobago was the most under-populated island in the British West Indies by 1946 (Library of Congress, 2006). Sugar production increased in Trinidad up to the 1960s, followed by decline; only a single sugar factory remained by the 1980s. The state sugar company closed in 2003, and the sugar factory in April 2010. Former cane areas have been neglected or developed for other purposes (e.g., housing). Thus although the population continued to increase, to about 1.3 million in 2000, the decline of the sugar industry reduced the pressure for destruction of natural habitats as former cane areas were utilized instead. Trinidad and Tobago has the lowest population growth rate of any of the territories considered in this series, at −0.87% per year (Wilson et al., 2006, 2011), due to emigration.

Apart from sugar, there has also been extensive cultivation of cacao at lower elevations in the Northern and Central Ranges, and in Tobago, for over 250 years. Cocoa was the main export between 1870 and 1930, and Trinidad and Tobago was the second largest producer in the world (Watts, 1987; Library of Congress, 2006). Many cacao estates were abandoned after 1930, and remain so today (Van den Eynden, 2007). Cacao is usually an understory crop and retains considerable herpetofauna and other biodiversity (Faria et al., 2007), particularly when abandoned, and so is less environmentally damaging than sugar. Rice has been cultivated extensively around both the Caroni and the Nariva swamps, controversially in the case of the latter as swamp forest has been destroyed in the process. The Nariva Reforestation Project is being funded by the World Bank to reforest 1300 ha of illegal rice fields over the next 5 years.

Trinidad and Tobago became an independent nation in 1962, and a republic in 1976. The islands retain considerable extent and variety of wild habitats, and no ecosystem types have been lost (Kenny, 2008). About 51% of the land is forested, two thirds of which is utilized commercially, but only 5% is plantations (mostly teak and Caribbean pine; Chalmers, 2002). The rate of forest loss is about 0.8% per year. 15% of the land area is arable, and another 15% is built (MPUE, 2006). Economic development in the 20th century has largely been fuelled (metaphorically and literally) by the oil and gas industries; agriculture accounts for only about 6% of exports (Pemberton et al., 2002). Oil and gas production are associated with transport (pipeline) and refining activities, and "downstream" industries. As of 2002, Trinidad was the world's largest exporter of both ammonia and methanol (Encyclopedia of the Nations, 2010), and there are controversial plans for large scale aluminium smelting (Kenny, 2006). Tobago has a more typical Caribbean island economy largely based on tourism.

Conservation legislation

Biological conservation in the Republic of Trinidad and Tobago is the joint responsibility of two agencies. The Environmental Management Authority (EMA) drafts legislation (table 4a) and checks compliance with international treaty obligations (table 4b), following its National Environmental Policy (EMA, 2005). The Forestry Division regulates use of natural areas, and its Wildlife Section issues permits for hunting and collecting. Both agencies cover both Trinidad and Tobago, but the Department of Natural Resources and the Environment of the Tobago House of Assembly also has considerable powers. A permit has recently (2005) been introduced to licence all biodiversity studies in Tobago, after application and payment of a US$100 fee, a system expected to be adopted in Trinidad.

Currently only sea turtles (when on land) and some reptiles receive protection under the Conservation of Wild Life Act (CWLA) (table 4), which defines "animals" as any mammal, bird or reptile (thus excluding amphibians). Schedule 2 (game animals) includes caiman, iguana, and matte (*Tupinambis*), which may be hunted under licence from October to February (formerly to March). Schedule 3 (vermin) includes venomous snakes, specifically the mapipire balsin (*Bothrops*), mapipire zanana (*Lachesis*), and coral snakes (*Micrurus*), which may be destroyed by the landowner at any time. Other "animals" not listed on Schedules 2 or 3 are protected, but may be captured for scientific purposes under a Special Game Licence. Cooper (2008) illustrated lizards collected for a foreign museum, but confiscated in Trinidad for lacking this licence. Hunting requires an annual State Game Licence, issued by

Table 4. (a) National environmental legislation (and most recent amendment) and (b) international treaty obligations relevant to herpetofauna conservation in the Republic of Trinidad and Tobago (MLA, 2004; Bräutigam and Eckert, 2006).

National Legislation	Year	Chapter	Purpose
Environmental Management Act	2000	35:05	Establishes the EMA, to promote understanding and conservation of the environment and to develop a National Environmental Policy.
Environmentally Sensitive Species Rules	2001	Subsidiary LN 63	Prohibit taking, possession or trade of an organism or its parts, removal or export, or other activities likely to cause harm to the species.
Environmentally Sensitive Areas Rules	2001	Subsidiary LN 64	Protect a defined area, including prohibiting removal of or damage to all animals in the area.
Institute of Marine Affairs Act	1976, 1996	37:01	Establishes the IMA to advance marine scientific research and advise Government on marine matters.
Marine Areas (Preservation and Enhancement) Act	1970, 1996	37:02	Creation of restricted areas to protect flora and fauna, natural beauty, public enjoyment, or promote scientific study and research.
Forests Act	1915, 1999	66:01	Regulates use of forests and forest produce.
Forests (Prohibited Areas) Order	1953, 1990	Subsidiary	Restricts entry to Matura and Fishing Pond beaches.
	1997	GN 125	Restricts entry to Grande Riviere beach.
Conservation of Wild Life Act	1958, 1980	67:01	Regulates hunting and scientific collecting, export of animals, Game Sanctuaries.
Animals (Diseases and Importation) Act	1954, 1997	67:02	Regulates import of animals and requires notification of animal diseases.
Fisheries Act	1916, 1975	67:51	Regulates fishing methods and species, including sea turtles.
Mongoose Act	1918	67:55	Bans import or keeping, allows ordering destruction by landowners.

Table 4. (b).

International Treaty	Adoption in Trinidad and Tobago	Purpose
Convention on International Trade in Endangered Species of Wild Fauna and Flora (CITES)	Entered into force 1984	Regulates import and export of wild life.
Convention on Wetlands of International Importance (Ramsar)	Entered into force 1993	Conservation of wetlands.
Convention on Biological Diversity (CBD)	Ratified 1996	Protection of biodiversity.
Cartagena Convention, Protocol Concerning Specially Protected Areas and Wildlife (SPAW)	Ratified 1999	Protection of marine biodiversity, including sea turtles.

the Wildlife Section at six centres in Trinidad (fig. 2a) and in Tobago, whereas the Special Game Licence for research is issued at the head office in St. Joseph. The State Game Licence is accompanied by up to five optional permits (TT$20 each; TT$6.3 = US$1) out of eight categories of game. The "alligator" permit allows capture of unlimited numbers of caiman, iguana, and matte; the number and total weight (in lbs) caught on each permit are later self-reported on an annual hunting return. Special Game Licences are also issued for amphibians, although this is only a legal requirement if specimens are to be exported under CITES regulations. A new Conservation of Wildlife Bill was drafted in 2003 (EMA, 2004, 2006), but this has not been enacted. Caiman, iguana and matte were on Schedule 1 (species requiring a harvesting licence), and the five sea turtles and the tortoise *C. denticulata* on Schedule 4 (protected species harvested only under a special licence, e.g., for scientific research). Consideration of the turtle fishery probably prevented enactment of this bill, as occurred with the second series of Environmentally Sensitive Species (below).

Sea turtles have an ambiguous legal status in Trinidad and Tobago. They are protected under the CWLA, but a legal harvest is permitted under the Fisheries Act, which defines "fish" to include turtles. In practice, turtles on land, their eggs, and female turtles within 1000 yards of the shore or within reefs, are absolutely protected by law. Outside these limits, all sea turtles may be legally caught and sold, apart from during a closed season from March to September. The three most important leatherback nesting beaches, a total of almost 20 km, have been declared as prohibited areas under the Forests (Prohibited Areas) Order (table 4). A permit from the Wildlife Section is required to visit these beaches, and permit holders must be accompanied by an accredited tour guide.

The legislation creating National Parks in Trinidad and Tobago has not been completed, so that although some exist on paper (such as Matura) they do not receive effective protection. There are also 13 Wildlife Sanctuaries and 35 Forest Reserves (MPUE, 2006) which are similarly ineffective (EMA, 2001). The inadequacy of legal protection of environment and biodiversity in Trinidad and Tobago led to the

Environmental Management Act of 1995 (re-enacted in 2000 due to a technicality) which established the EMA, and the Environmentally Sensitive Areas (ESA) and Environmentally Sensitive Species (ESS) Rules (table 4). The ESS Rules prohibit removal of an organism (except for scientific research approved by the relevant authority), disturbance of the habitat of the species or of the organisms themselves, pollution of the habitat, or any other activity likely to cause harm to the species. The ESA Notices (defining particular areas) define "animal" as any member of the animal kingdom, and prohibit removal of or damage to animals in the area. Amphibians living within an ESA are thus protected.

The first series of three ESSs (the manatee, Trinidad piping-guan, and sabrewing hummingbird) was selected internally by the EMA and designated in 2005. The second series selected comprised the five species of sea turtles, but their designation was rejected in Cabinet, apparently due to pressure from members concerned about closure of the turtle fishery. The third series was selected by the EMA after a consultation exercise, because there was a wide range of candidate species; the process is of interest as herpetofauna received substantial support, perhaps against expectation. External experts were invited to summarize the conservation problems of particular taxa at a stakeholder consultation in April 2006: plants, corals and molluscs, marine fish, amphibians and reptiles, birds, and mammals. The consultation was attended by invited representatives of community-based conservation groups, naturalists, eco-tour operators, hunting groups and others. Attendees scored three species in each taxonomic group for ten criteria: as a flagship, umbrella, biodiversity indicator, sentinel, or keystone species, and for endangerment, immediacy of threat, utility, uniqueness, and research potential. The top scoring seven species of each attendee were then awarded 1-7 ranking points. The top ranking six species overall were the ocelot (162 points), *Phytotriades auratus* (122), stony corals (94), silky anteater (90), *Mannophryne olmonae* (75), and orchids (57). After internal review, the ocelot and golden tree frog were selected by the EMA as the proposed third series of ESS, and the draft Notices are currently (April 2010) open for public comment before being considered by Government. It is notable that the golden tree frog was selected in preference to other species with a much greater local conservation profile such as the scarlet ibis.

Environmentally Sensitive Areas are in many cases the same as the National Parks, although the boundaries do not coincide exactly. The first three were Matura (fig. 2a), a forested area at the eastern end of the Northern Range (designated in 2004); Nariva Swamp (2005); and Aripo Savanna, an edaphic wet grassland with forest islands (2007). Several other ESAs have been proposed, including Caroni Swamp, Trinity Hills, Buccoo Reef, and Main Ridge Reserve (fig. 2b). The south-west peninsula of Trinidad and the north-east of Tobago, both of which are important for herpetological diversity (see distribution maps in Murphy, 1997) are the main regions where additional protected areas are needed to conserve amphibians and reptiles.

The Animals (Diseases and Importation) Act (table 4) defines "animal" as all animals of whatsoever kind, and so implicitly includes amphibians. Reptiles are explicitly included, with no bird, reptile or insect allowed to be imported except under licence. This Act thus potentially allows some control of species introductions, and also deals with notification of diseased animals and could be used in actions against chytridiomycosis. The Marine Areas (Preservation and Enhancement) Act enables the creation of restricted areas to protect flora and fauna, or to promote scientific study and research, and restricts entry except with permit, and could be used to protect areas important for sea turtles. The Institute of Marine Affairs Act established the IMA to advance marine scientific research and advise Government, and the IMA has been active in research and public education on sea turtles.

Hunting and utilization

There is much hunting for sport and food in Trinidad and Tobago, including of caiman, iguana (fig. 3) and *Tupinambis*. Caiman were also previously used in the curio trade (Kenny, 2008). The number of hunters in Trinidad and Tobago has increased from about 3500 in 1975 (Cooper and Bacon, 1981) to about 10,000 today (R. Sorrillo, pers. comm., 26 April 2010). Hunting is less prevalent in Tobago; only about 60 permits were issued per year in the late 1990s (EMA, 2001), compared to about 8000 in Trinidad. Van den Eynden (2007) and local assistants interviewed hunters living in and visiting the Matura National Park, and found that 21% of the rural population hunted, for an average of 38 days per year.

There has been no full analysis of hunting returns, most of which were recently contaminated in storage and have since been destroyed. Returns from Trinidad for the 2000/2001 season were digitized by the Department of Food Production of the University of the West Indies and are analysed here. These records show 5939 hunters, who purchased 8453 permits (1.42 permits per hunter). Only 387 permits were for "alligator"; 4.6% of all permits issued, to 6.5% of hunters. There was significant variation in the proportion of permits for "alligator" among the six offices (χ^2 test, $P < 0.001$), with a very low value at Rio Claro (0.8% of permits), the reason for which is obscure. 160 hunters submitted non-zero returns for the "alligator" permit, accounting for 1430 animals caught, a mean of 8.9 per hunter; there was no significant difference among the six offices (Kruskal-Wallis test, $P = 0.64$). Most returns were for 1-10 animals, but 16 hunters claimed more than 20 catches, the highest being 73 (fig. 4a). Van den Eynden (2007) recorded iguanas as the only reptiles hunted in the Matura National Park; their value was TT$60-100 each (compared to, e.g., TT$100-300 for agouti, the most popular quarry with permits taken out by 55% of hunters). It is probable that iguanas make up the large majority of catches under the "alligator" permit throughout Trinidad, the modal size of animal caught being 1-2 kg (fig. 4b); animals above 4 kg were probably *Caiman*. There was no significant difference in mean size of animals caught among the six offices (Kruskal-Wallis test, $P = 0.16$). Van den Eynden (2007) found that less than

Figure 3. Trinidadian hunter with a large *Iguana iguana* (total length c. 1.8 m, mass 3-4 kg). Photo by Kevin Mahabir, 2007 (Colour original — see www.ahailey.f9.co.uk/appliedherpetology/cariherp.htm).

half of hunters interviewed reported catches under permit to the Wildlife Section, and unlicensed hunting is also prevalent; perhaps particularly for iguanas, which are often taken opportunistically, including in suburban areas.

Hunting of mammals and birds is also likely to have some negative effects on herpetofauna, most directly through presence of hunters and their dogs in the forest, killing snakes. Snake bites are a regular risk to hunting dogs, which their owners treat using traditional remedies (Lans et al., 2001). Hunting dogs are valuable, and most hunters seem to take every opportunity to kill any snakes encountered, venomous or not (Van den Eynden, 2007). Persecution of venomous snakes is widespread in Trinidad, although fatal bites are infrequent. A recent case, of unknown species, was reported as the first in several years (John, 2009). This took place in a yard, not in the forest. There may also be indirect negative effects of hunting of other taxa on herpetofauna; Zimmerman and Bierregaard (1986) found

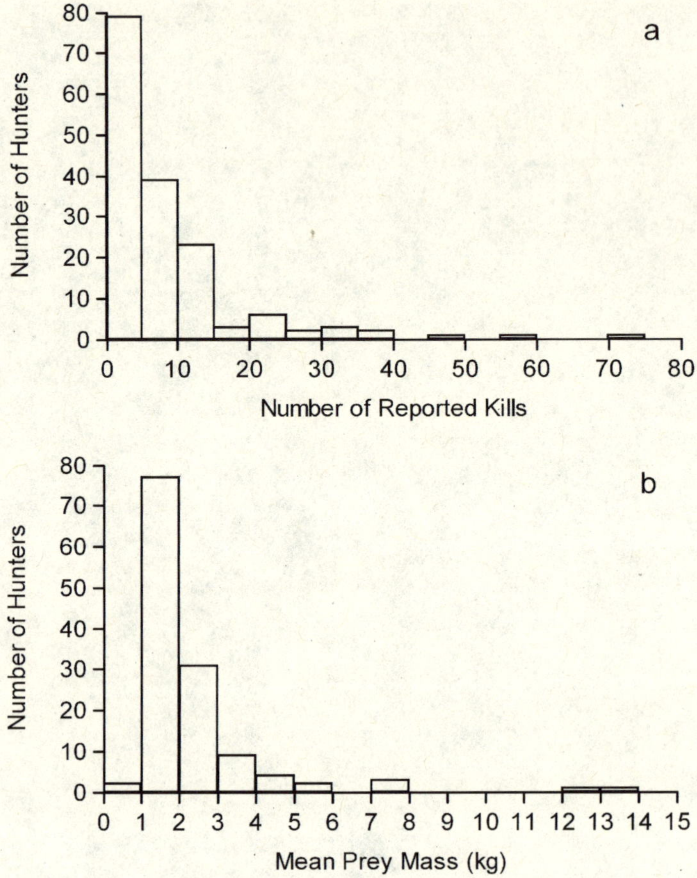

Figure 4. Results from Trinidadian hunters reporting on "alligator" permits (valid for *Iguana*, *Tupinambis* and *Caiman*). (a) Frequency distribution of number of animals caught (excluding zero returns), $n = 160$. (b) Frequency distribution of mean animal size (kg), $n = 130$.

that Amazonian frogs had declined in forest fragments without peccaries to provide breeding pools.

Tortoises were formerly hunted, being seen in markets up to the 1950s; *C. denticulata* was widespread and considered a delicacy, though it is now rarely seen (Kenny, 2008). There is evidence that it used to occur in Tobago, together with a large extinct edible frog, both possibly extirpated by hunting in Amerindian or early colonial times (Hardy, 1982). Cooper and Bacon (1981, Appendix II) used the name "mountain chicken" for the large frog *Leptodactylus bolivianus*, apparently by analogy with *L. fallax* of Dominica and Montserrat. The name is not used popularly, and there is no evidence of current utilization of *L. bolivianus*, which is a relatively uncommon forest-edge frog. As noted by J.S. Kenny (pers. comm., 17 October 2003), "no-one in Trinidad eats frogs". *Iguana* eggs, *Boa* fat, and whole young

Bothrops and *Lachesis* may be used in traditional medicine in Trinidad (Van den Eynden, 2007), and turtle eggs and organs are said to be used as aphrodisiacs.

There is a small artisanal turtle fishery in Trinidad and Tobago; historical data are given by Rebel (1974). The fishery is based on hard-shelled turtles (i.e., not leatherbacks), using nets of 15-30 cm mesh 30 m long (which may be joined in series), suspended from the surface to a depth of 2-3 m. Nets are set at known foraging areas of the turtles, and checked each morning and evening. Most turtles are caught at night. Harpoons may also be used, especially at Grande Riviere, Toco and Carenage (Lee Lum, 2003). The marine turtle fishery is principally of green turtles and hawksbills, making up 50% and 47.5% of the total in Trinidad (Chu Cheong, 1995), the remainder being loggerheads. The species composition of the catch depends on the area fished; green turtles in seagrass beds, and hawksbills in rocky areas and reefs.

Chu Cheong (1995) of the IMA surveyed 15 fishing depots in Trinidad from 1982-1983 and found turtle fishing at six depots. Only 1-4 persons were involved at each depot (a total of 12 persons), as a part-time activity. The same investigator (Lee Lum, 2003) surveyed 27 depots in 2001-2002 and found turtle fishing at eight depots, also with a total of 12 persons involved. Only four depots had turtle fisheries in both surveys (Matelot, Toco, Mayaro, and La Lune). The sea turtle fishery in Trinidad is thus small scale, with an admitted 4-10 turtles caught per week at each depot in 1982-1983. The changing incidence of turtle fishing at depots suggests that the fishery is opportunistic rather than traditional in any one location. Data are available for the weight and value of turtle meat sold up to 1980, but since then no data have been collected by the Fisheries Division, because the redesigned fish reporting form did not include turtles (Chu Cheong, 1995).

The IMA survey did not cover Tobago, but turtle fishery statistics from before 1980 (in Chu Cheong, 1995) show that quantities landed were only a small fraction of those at Trinidad depots (18-250 kg per depot per year, compared to, e.g., 5245 kg per year at Carenage). In a survey of 215 fishermen around Tobago in 2007 (Cazabon-Mannette, unpubl. data), 22 confirmed that they targeted turtles, and 14 considered turtles an important source of income. One who fished regularly during the open season reported catching 150 turtles per year. These 22 fishermen were interviewed at landing sites all around Tobago; turtle fishing activity was not restricted to the south-west as previously reported. The price of turtle meat was usually TT$25-30 per pound, about twice that obtained for fish.

Invasive competitors, predators and diseases

Invasive herpetofauna are likely to have little effect on native species as most are currently restricted to man-modified habitats, although *Hemidactylus mabouia* may have displaced *H. palaichthus* on the main islands. Among the introduced anoles, *A. wattsi* is the most ground-dwelling and would have the greatest effect on the native *A. chrysolepis* if it spreads into forested areas (White and Hailey, 2006). None of the introduced *Anolis* are likely to hybridize with *A.* (*Norops*) *chrysolepis*, but hybridization occurs between *A. aeneus* and *A. trinitatis* and may be partly responsible for the decline of the latter species (Gorman et al., 1971; Hailey et al., 2009). *Eleutherodactylus johnstonei* has been implicated in declines of frogs in other islands (Henderson and Berg, 2006, 2011); the most likely native (and endemic) species to be affected is *Pristimantis* (formerly *Eleutherodactylus*) *urichi*. The nuisance of the loud calling of *E. johnstonei* has already been felt, which can only exacerbate the largely negative local opinions of herpetofauna. More importantly, there could be acoustic interference with other frogs (Hedges, 1993), whose calls have evolved in the absence of this species.

Invasive and introduced non-herpetofauna are more problematic. The small Indian mongoose was introduced at Santa Cruz in the Northern Range in 1875 (Lever, 1985), to control rats and/or snakes. It spread rapidly, already being considered a pest by 1902 when a bounty was introduced (Williams, 1918). The mongoose became common in agricultural and abandoned land, but not in natural forests. It is absent from Tobago (Nellis and Everard, 1983). The preferred habitat seems to be dry bush, as in other West Indian islands or in India; perhaps native mammal predators also exclude the mongoose from forest in Trinidad. Bounty records show that the mongoose declined after 1930, for an unknown reason; Nellis and Everard (1983) state that Trinidad was among the islands most heavily damaged, but is now the least affected, by the mongoose which is now at ecological equilibrium at low population density. Nevertheless, it is likely that the herpetofauna has been affected; Urich (1931) noted the scarcity of *Ameiva*, small ground snakes, *Rhinella* and *Leptodactylus* in areas with mongooses. Williams' (1918) study of stomach contents estimated that an individual would consume 17 lizards, 18 snakes, and 29 anurans in a 3-month period, with frogs being taken particularly in sugar cane fields, and snakes in cacao plantations.

The red imported fire ant *Solenopsis invicta* was observed near the Caroni Swamp in 2000 (Davis et al., 2001), and has spread to urban areas in Trinidad and to Tobago (J.K. Wetterer, pers. comm., 16 October 2004); it is particularly common in sugar cane areas (Wetterer and Davis, 2010). The fire ant is a threat to eggs, hatchlings, and adults of terrestrial reptiles (Wetterer and Moore, 2005), and to eggs and hatchling sea turtles (Wetterer et al., 2007). A synergistic action is possible between any beach loss from global warming and presence of the fire ant, since ants increase higher up the beach closer to dune vegetation. Beach replenishment which widens a beach thus reduces ant predation (Wetterer et al., 2007), but reduced beach width from sea level rise and erosion might increase exposure to ants.

Potentially the most damaging invasive species to the herpetofauna is the chytrid fungus *Batrachochytrium dendrobatidis*, which has been implicated in amphibian declines worldwide. The fungus was detected at high prevalence (25%) using DNA analysis of skin swabs of *Mannophryne olmonae* in Tobago in 2006 (Alemu et al. 2008). The fungus has also been detected using the same methods in Trinidad, with two positive results in 2007 and a further eight in 2009, all from two populations of *M. trinitatis*; those at highest altitude among 12 populations examined (J.B. Alemu

I et al., unpubl. data). *Mannophryne* species appear to be resistant carriers of the amphibian chytrid, and do not show symptoms of chytridiomycosis; other local species may not be so fortunate.

Incidental catch of sea turtles

The bycatch of sea turtles is a large problem in Trinidad and Tobago waters, causing more deaths of leatherbacks than all others causes combined (Eckert and Eckert, 2005). Leatherbacks are principally caught in gillnets set for kingfish and carite, when the turtles approach nesting beaches. Lee Lum (2006) estimated that more than 3,000 leatherbacks were caught by the gillnet fishery in Trinidad in 2000, mostly on the north and east coasts from January to August when turtles were arriving to nest. Hawksbill and green turtle incidental catches were reported to be rare, but these species are clandestinely eaten when captured, even in the closed season (Lee Lum, 2003). Pritchard (1984) noted that the incidental catch of the rare olive ridley turtle might be proportionately important, as this slow-swimming species is easily caught in trawls.

About 67% of leatherback turtles were released alive after incidental catch, usually requiring nets to be cut (Fournillier and Eckert, 1998; Lee Lum, 2006). A grant for net repairs was offered at one time under an UNEP-funded programme, but is no longer given. This compensation was effective in reducing turtle mortality as fishers were more amenable to cutting their nets to release turtles quickly (and alive). Incidental catches are associated with green multifilament nets, as these are used at night (when turtles approach nesting beaches) at the surface, and are strong enough to hold turtles. In contrast, nylon monofilament nets are set in the day, anchored to the sea floor, and rarely catch turtles, and are not sufficiently strong to hold leatherbacks if caught (Lee Lum, 2003). Incidental catches in gillnets are mostly within 5 km of the beach, although these nets are used up to 15 km offshore. Incidental capture was particularly a problem on the north coast, where turtles were described as "thick in the water" in the peak nesting season. Nevertheless, all fishers surveyed by Lee Lum (2003) knew of the regulations concerning sea turtles, and 65% agreed with them.

Turtles and tourism

A number of activities associated with the large amount of general tourism development in Tobago are detrimental to sea turtles. These include (Fournillier and Eckert, 1998): sand mining and beach erosion; hotel development (often unlicensed); release of sewage offshore; photopollution from beachfront lighting affecting adults and hatchlings; disturbance from tourists approaching nesting females; mechanized beach cleaning; presence of predatory feral dogs; retention of hatchlings by hotel management; boat mooring and diving on reefs; sedimentation of reefs and seagrass beds.

On the other hand, ecotourism is an increasing economic activity and factor in conservation in Trinidad and Tobago. The Main Ridge Rain Forest won awards as the World's Leading Ecotourism Destination annually from 2003-2006, and as the World's Leading Green Destination in 2007 and 2009 (WTA, 2010). Much of the current focus of local ecotourism is marine, including sea turtles. There are many community-based conservation and ecotourism groups, particularly on the north and east coasts of Trinidad (table 5); the most well-known is Nature Seekers at Matura. Ecotourism has been shown to positively affect local attitudes to turtles in Trinidad (Waylen et al., 2009). The Turtle Village Trust (TVT) was formed in 2006 as the umbrella organisation for local turtle conservation organisations, and aspires to foster partnerships among community groups, government and corporate entities "to place Trinidad and Tobago as the premier turtle tourism destination globally". Some 46,000 visits were recorded at beaches to watch turtles nest in 2009; about 30,000 at Grande Riviere and the balance at Matura (fig. 5a). Locals made up the majority of visitors, but over 13,000 were foreigners. TVT has increased and standardized survey efforts and coordinated tagging across important nesting beaches to improve population estimates, and installed signs in communities across north-east Trinidad (fig. 5b). Public awareness and capacity building are also areas of focus; TVT staff make presentations at schools and events, produce informative brochures and flyers, and are involved in training members of the communities.

The Future — Threats and Recommendations

Two additional threats to herpetofauna have not been studied in Trinidad and Tobago but are likely to have increasing impact. These are threats that will occur as a result of future development of the islands, and the local effects of global climate change.

The Government of the Republic of Trinidad and Tobago previously (to May 2010) had as its major policy the achievement of developed country status by 2020 (the Vision 2020 programme). Development will bring increasing threats of habitat loss and pollution from industrialization, urbanization, transport links, and quarrying (which is often illegal in Trinidad, and involves much environmental damage through siltation of rivers, in addition to the habitat removed). The current (June 2010) Government does not stress the development of industry to the same extent, but its focus on agriculture (which already uses large quantities of pesticides in Trinidad — Pemberton et al., 2002) also has potential for environmental damage. Indirect effects of pesticides on the herpetofauna (e.g., on prey populations) could also be important. Work on these threats in Trinidad and Tobago is in the early stages, e.g., finding a suitable local model for ecotoxicological studies (Hailey et al., 2006, 2007). Salinization from produced water from the oil industry is also a threat to amphibians. Currently 17 inland wells have total produced water discharge rates of 100,000 barrels per day, with a salinity of 8-33 ppt, 90% of which enters waterways (Elias-Samlalsingh and Agard, 2004). Seismic surveys for oil and gas are potentially a threat to sea turtles. Increased utilization of forest resources is

Table 5. Community-based and coordinating conservation groups in Trinidad and Tobago involved with sea turtles.

Group	Location	Associations
Blanchiseusse Environment and Art Trust (BEAT)	Blanchiseusse	
PAWI[1] Sports, Culture and Eco Club	Matelot	UNDP; Caribbean Forest Conservation Association; Wildlife Section; University of Glasgow (Livingstone and Downie, 2005).
Grande Riviere Nature Tour Guide Association (GRNTGA)[2]	Grande Riviere	Wildlife Section; Institute of Marine Affairs (Peters, 2004; Lee Lum, 2005).
Sans Souci Wildlife Tours	Sans Souci (Toco)	Toco Foundation.
Toco Foundation	Toco	Sans Souci Wildlife Tours.
Nature Seekers	Matura	Wildlife Section; WIDECAST, Duke University.
Fishing Pond Environmental Community Group (FPECG)	Fishing Pond	Wildlife Section.
Fishing Pond Turtle Conservation Group (FPTCG)	Fishing Pond	Wildlife Section.
Manatee Conservation Trust	Manzanilla	Wildlife Section.
Manzanilla Wildlife and Environmental Project (MWEP)	Manzanilla	
Nariva Environment Trust	Manzanilla	Wildlife Section.
Save Our Sea Turtles Tobago (SOSTobago)	Tobago	Department of Natural Resources and the Environment; Barbados Sea Turtle Project (Clovis, 2004).
Turtle Village Trust	Trinidad and Tobago	Wildlife Section; Department of Natural Resources and the Environment; Ministry of Tourism; BHP Billiton; Atlantic LNG; IFAW; Community groups.

[1] The local name of the Trinidad piping-guan, not an acronym.
[2] Formed in 2001 as the successor to the Grande Riviere Environmental Awareness Trust (GREAT).

Figure 5. (a) Turtle nesting tour at Matura (photo courtesy Nature Seekers) — the lamp has a red screen. This group is mostly local visitors. (b) Turtle Village Trust welcome sign at Grande Riviere, where there is also a visitor centre (Colour originals — see www.ahailey.f9.co.uk/appliedherpetology/cariherp.htm).

also probable. Local forestry practices have been shown to affect biodiversity of bats relatively little (Clarke et al., 2005); thermal and other possible effects on herpetofauna (e.g., Vitt et al., 1998) await study.

Global climate change is expected to have major effects on the economy and environment of Small Island Developing States (Mimura et al., 2007). Higher sea levels might reduce the size of sea turtle nesting beaches (Fish et al., 2005), increase the risk of erosion of nests (Lee Lum, 2005), and bring nests closer to dangers from invasive fire ants. Rising sea levels may also cause salinity increases in coastal fresh water, such as swamps, ditches and temporary pools used by breeding amphibians. Tropical herpetofauna experience stable climates, and may be unable to adapt to change even though temperature increases in the tropics are predicted to be smaller than those at high latitudes. Lowland tropical lizards are thus vulnerable as they

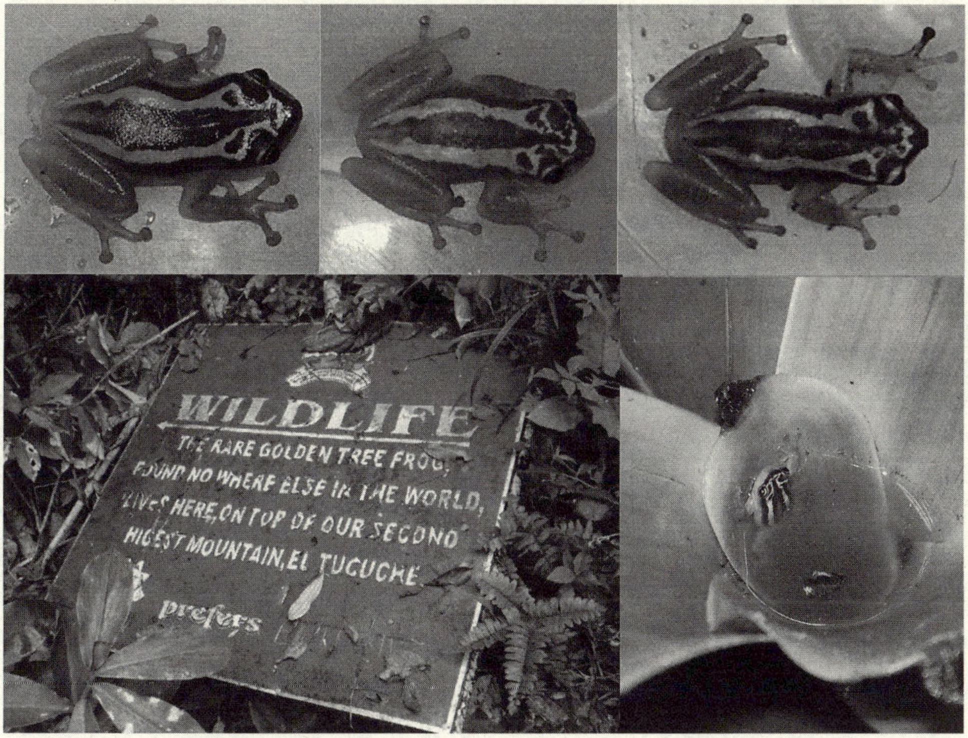

Figure 6. The golden tree frog *Phytotriades auratus*, a Critically Endangered Trinidad endemic. Captive frogs, showing variation in markings sufficient for individual recognition in captivity. Fallen conservation sign on El Tucuche, February 2010, these signs were placed high in the hills, where they were hard to maintain (and possibly less useful). Captive frog in tank of bromeliad, *Glomeropitcairnia erectiflora*. (Colour originals — see www.ahailey.f9.co.uk/appliedherpetology/cariherp.htm).

already exist close to their thermal maxima, and would be under physiological stress from even a small rise in temperature (Huey et al., 2009). The lowland herpetofauna of Trinidad and Tobago is perhaps of lesser conservation importance, as it is similar to the mainland fauna; endemic species in the highlands are of much greater concern. Many species may simply migrate upwards with increasing temperatures as competition from warm-adapted species increases from below, but this option is not available to species found only at high elevations, hence the importance of the golden tree frog in local conservation (fig. 6). Lena Dempewolf (pers. comm.) and others are currently assessing the geographic and exact elevation limits of the frog and the giant bromeliad. The generally accepted relation between temperature and elevation on mountains is a decrease of 1°C for an ascent of 150 m (MacArthur, 1972). If *P. auratus*, or its bromeliad, is sensitive to high temperatures, then their current elevation limit suggests that they will be vulnerable to any increase of mean temperature over 1°C. The target to limit global warming to 2°C would thus condemn *P. auratus* to extinction, even if practical and legal protection measures in progress are successful. Global warming may also increase the risk of chytridiomycosis at montane tropical sites as the warmer climate becomes more suitable for the fungus (Pounds et al., 2006).

If global temperatures increase as predicted, the only option for protecting *P. auratus* would be ex-situ conservation, either in captivity or translocation to higher elevation habitats outside of Trinidad, e.g., to Venezuela where *G. erectiflora* also occurs (Clarke et al., 1995). There have been approaches from Chester and Paignton zoos (UK) to set up captive breeding populations of *P. auratus* (R. Gibson and M. Bungard, pers. comm., January 2010). Preliminary study of three captive individuals (A. Hailey, pers. obs.) has shown that both frogs and bromeliads may be maintained successfully, and their markings are sufficient for individual recognition in a small group (fig. 6). Locating frogs within bromeliads has proved difficult; two were later found in a 40-cm tall bromeliad examined carefully on El Tucuche and brought down for culture, and one hid in a 15-cm bromeliad moved between cages. Non-destructive searching of bromeliads is thus likely to underestimate wild populations. Collection and destructive examination of bromeliads unfortunately still occur, but would be prohibited if *P. auratus* becomes an ESS, with restrictions on disturbance of its habitat.

Recommendations

1. Specific legal protection for all endemic herpetofauna. The concept of the ESS is for a small number of flagship species, and is unlikely to be extended to other amphibians or reptiles. An updated version of the draft Conservation of Wildlife Bill should therefore include all endemic species on Schedule 4 (protected species harvested only under a special licence).
2. Recognition of amphibians as "animals" in the legislation replacing the CWLA, so that they receive the same protection as other non-game and non-vermin species. The ignoring of amphibians seems to be a general pattern in laws of the former British West Indies territories, presumably reflecting their similar legislative backgrounds. Attendees from several islands at a recent workshop on chytridiomycosis in the Caribbean (Dominica, March 2008) noted this pattern, with amphibians only being recognized in law in those territories with game species (mountain chickens, in Dominica and Montserrat).
3. Inventory the herpetofauna in all existing and proposed protected areas to show whether those are deficient for any threatened species, and if so to expand the network, e.g., in the south-west peninsula of Trinidad and north-east Tobago. In addition, the protection offered to non-game animals by Wildlife Sanctuaries and Forest Reserves should be increased.
4. Research on herpetofauna populations and species diversity in key habitats, and the effects of habitat degradation, forestry, pollution and environmental change on local herpetofauna species and communities.
5. Restrict the spread of the amphibian chytrid fungus using approved hygienic methods such as those discussed at the Dominica chytridiomycosis workshop

(Cunningham, 2006). Spread of this pathogen into, and then between, the high elevation areas occupied by the golden tree frog is particularly to be avoided.

6. Research on alternative conservation measures such as habitat creation or regeneration, for example of montane forests and bromeliads, and translocation of herpetofauna to form new populations. This may be the most effective short-term strategy to conserve the golden tree frog, though ineffective against global warming.
7. Development of captive populations of the golden tree frog, both in Trinidad and overseas, as an insurance against unsuitability of montane areas after global warming, and against chytridiomycosis in wild populations.
8. Improved data gathering on hunting of lizards and caiman, and on their population levels. At least half of all legal hunting is unreported, and there is in addition much poaching, including hunting in the closed season. There is currently no basis on which to determine whether hunting levels of reptiles (or any other game) are sustainable.
9. Full legal protection of sea turtles, and restriction of access at night to all important nesting beaches.
10. Practical protection of sea turtles at nesting beaches by additional funded patrols. Existing funded and voluntary turtle patrols (e.g., by students of the University of the West Indies Biological Society) cannot cover more than a small fraction of the total places and times required for effective protection of nesting turtles.
11. Increased use of international volunteer groups to patrol turtle nesting beaches, adding to successful programmes such as Earthwatch at Matura and the University of Glasgow on Trinidad north coast and Tobago index beaches.
12. Replacement of the turtle fishery by alternative income sources for fishers, such as their involvement on boats in the lucrative dive ecotourism sector (in Tobago; Trinidad waters are too opaque). Until this can be achieved, turtles should be added to the Fisheries Division fish reporting form so that the scale of the catch can be monitored.
13. Reduction of the incidental catch of leatherback turtles using fishery regulation, such as prohibited areas around the main nesting beaches in north-east Trinidad or restrictions on the depth at which nets are set, or practical turtle-repellent devices.
14. Increase local capacity through training and employment of herpetologists with environmental agencies, and increased availability and use of local information sources and reference collections in education to overcome negative local attitudes to herpetofauna.
15. Finally, it should be recognized that development has benefits as well as risks for conservation. Developed country status should be associated with high standards of wildlife protection and conservation, and compliance with international obligations.

Acknowledgements. We thank Angela Ramsey, Dave Boodoo, Nadra Nathai-Gyan, and Richard Sorrillo for help of the Department of Natural Resources and the Environment and the Wildlife Section; Mike Bungard, Richard Gibson, Blair Hedges, Julian Kenny, Michael Patrikeev, Basil Reid and Graham White for information; Kevin Mahabir for his photograph and Ryan Mannette for constructing the maps; and Roger Downie, Julia Horrocks, Julian Kenny, and Suzanne Livingstone for helpful comments on the ms. Our work on sea turtles was supported by grants from the Government of Trinidad and Tobago Research Development Fund and the National Fish and Wildlife Foundation (Washington).

References

Alemu I, J.B., Cazabon, M.N.E., Dempewolf, L., Hailey, A., Lehtinen, R.M., Mannette, R.P., Naranjit, K.T., Roach, A.C.J. (2008): Presence of the chytrid fungus *Batrachochytrium dendrobatidis* in populations of the Critically Endangered frog *Mannophryne olmonae* in Tobago, West Indies. EcoHealth **5**: 34-39.

Bacon, P.R. (1970): Studies on the leatherback turtle, *Dermochelys coriacea* (L.), in Trinidad, West Indies. Biol. Conserv. **2**: 213-217.

Bacon, P.R. (1981): The Status of Sea Turtle Stocks Management in the Western Central Atlantic. WECAF Study No. 7.

Beard, J.S. (1944): The natural vegetation of the island of Tobago, British West Indies. Ecol. Monogr. **14**: 135-163.

Beard, J.S. (1946): The Natural Vegetation of Trinidad. Clarendon Press, Oxford.

Boomert, A. (2000): Trinidad, Tobago and the Lower Orinoco Interaction Sphere: An Archaeological/Ethnohistorical Study. Alkmar, The Netherlands, Cairi.

Boos, H.E.A. (2001): The Snakes of Trinidad and Tobago. College Station, Texas A&M University Press.

Bräutigam, A., Eckert, K.L. (2006): Turning the Tide: Exploitation, Trade and Management of Marine Turtles in the Lesser Antilles, Central America, Colombia and Venezuela. Cambridge, TRAFFIC International.

Carranza, S., Arnold, E.N. (2006): Systematics, biogeography, and evolution of *Hemidactylus* geckos (Reptilia: Gekkonidae) elucidated using mitochondrial DNA sequences. Mol. Phylogen. Evol. **38**: 531-545.

Chalmers, W.S. (2002): Managing forest resources. In: Natural Resource Management for Sustainable Development in the Caribbean, p. 219-276. Goodbody, I., Thomas-Hope, E., Eds, Kingston, Canoe Press of the University of the West Indies.

Charan, R. (2007): Bathers warned to avoid snakes at South beaches. Trinidad and Tobago Express, 29 September 2007.

Chu Cheong, L. (1995): Report on the Sea Turtle Project 1981-1983. Part 3. A Survey of the Sea Turtle Fishery of Trinidad. Chaguaramas, Institute of Marine Affairs.

CITES (2010): Appendices I, II and III. (www.cites.org/eng/app/appendices.shtml, accessed on 26 April 2010)

Clark, P.U., Mix, A.C. (2002): Ice sheets and sea level of the Last Glacial Maximum. Quatern. Sci. Rev. **21**: 1-7.

Clarke, F.M., Pio, D.V., Racey, P.A. (2005): A comparison of logging systems and bat diversity in the Neotropics. Conserv. Biol. **19**: 1194-1204.

Clarke, F.M., Ward, A.I., Downie, J.R. (1995): Factors affecting the distribution and status of the golden tree frog, *Phyllodytes auratus*, in Trinidad. Brit. Herpetol. Soc. Bull. (54): 3-9.

Clovis, T. (2004): The status of marine turtle conservation in Tobago. Living World, J. Trinidad Tobago Fld. Nat. Club, Suppl., 30-31.

Cooper, M.E. (2008): Forensics in herpetology — legal aspects. Appl. Herpetol. **5**: 319-338.

Cooper, St.G.C., Bacon, P.R., Eds. (1981): The Natural Resources of Trinidad and Tobago. London, Edward Arnold.

Cummins, C.P., Swan, M.J.S. (1995): Variation in reproductive characteristics of the stream frog *Colostethus trinitatis* on the island of Trinidad. J. Trop. Ecol. **11**: 603-618.

Cunningham, A.A., Ed. (2006): Draft Chytridiomycosis Management Plan for the Lesser Antilles Region: Minimising the Risk of Spread, and Mitigating the Effects, of Amphibian Chytridiomycosis. London, Institute of Zoology.

Davis, L.R., Jr, Vander Meer, R.K., Porter, S.D. (2001): Red imported fire ants expand their range across the West Indies. Florida Entomol. **84**: 735-736.

Downie, J.R., Robinson, E., Linklater-McLennan, R.J., Somerville, E., Kamenos, N. (2005): Are there costs to extended larval transport in the Trinidadian stream frog, *Mannophryne trinitatis* (Dendrobatidae)? J. Nat. Hist. **39**: 2023-2034.

Eckert, K.L. (2001): Status and distribution of the leatherback turtle, *Dermochelys coriacea*, in the wider Caribbean region. In: Marine Turtle Conservation in the Wider Caribbean Region, p. 24-31. Eckert, K.L., Abreu Grobois, F.A., Eds, St. Croix, U.S. Virgin Is, WIDECAST.

Eckert, S.A. (2006): High-use oceanic areas for Atlantic leatherback sea turtles (*Dermochelys coriacea*) as identified using satellite telemetered location and dive information. Marine Biol. **149**: 1257-1267.

Eckert, S.A., Eckert, K.L. (2005): Strategic Plan for Eliminating the Incidental Capture and Mortality of Leatherback Turtles in the Coastal Gillnet Fisheries of Trinidad and Tobago: Proceedings of a National Consultation. Technical Report No. 5. Beaufort, North Carolina, WIDECAST.

Elias-Samlalsingh, N., Agard, J.B.R. (2004): Application of toxicity identification evaluation procedures for characterizing produced water using the tropical mysid, *Metamysidopsis insularis*. Environ. Toxicol. Chem. **23**: 1194-1203.

EMA (2001): Biodiversity Strategy and Action Plan for Trinidad and Tobago. Port of Spain, Environmental Management Authority.

EMA (2004): Environmentally Sensitive Species (ESS): Convention Matrix for Priority Species. Port of Spain, Environmental Management Authority.

EMA (2005): National Environmental Policy. Port of Spain, Environmental Management Authority.

EMA (2006): Proposed Environmentally Sensitive Species (ESS). Port of Spain, Environmental Management Authority.

Emsley, M.G. (1966): The mimetic significance of *Erythrolamprus aesculapii ocellatus* Peters from Tobago. Evolution **20**: 663-664.

Encyclopedia of the Nations (2010): Trinidad and Tobago — Energy and Power (http://www.nationsencyclopedia.com/Americas/Trinidad-and-Tobago-ENERGY-AND-POWER.html, accessed on 10 May 2010).

Environment Tobago (2010): Environment TOBAGO: Tobago's First Conservation Organization (http://www.scsoft.de/et/et2.nsf, accessed on 4 May 2010).

Faria, D., Paciencia, M.L.B., Dixo, M., Laps, R.R., Baumgarten, J. (2007): Ferns, frogs, lizards, birds and bats in forest fragments and shade cacao plantations in two contrasting landscapes in the Atlantic forest, Brazil. Biodivers. Conserv. **16**: 2335-2357.

Fish, M.R., Côté, I.M., Gill, J.A., Jones, A.P., Renshoff, S., Watkinson, A.R. (2005): Predicting the impact of sea-level rise on Caribbean sea turtle nesting habitat. Conserv. Biol. **19**: 482-491.

Fournillier, K., Eckert, K.L. (1998, 2009): Draft WIDECAST Sea Turtle Recovery Action Plan for Trinidad and Tobago. CEP technical report No. 38. Kingston, Jamaica, UNEP Caribbean Environment Programme.

Frost, C.D., Snoke, A.W. (1989): Tobago, West Indies, a fragment of a Mesozoic oceanic island arc: Petrochemical evidence. J. Geol. Soc. Lond. **146**: 953-964.

Frost, D.R. (2010): Amphibian Species of the World: An Online Reference. Version 5.4 (8 April, 2010). American Museum of Natural History, New York, USA (http://research.amnh.org/vz/herpetology/amphibia/, accessed on 20 April 2010).

Gaskin, M.R., Shephard, K. (1994): Sea Turtles and Their Habitats, Trinidad and Tobago and the Caribbean. Trinidad, Pointe-a-Pierre Wild Fowl Trust.

Gavilan, F.M. (2001): Status and distribution of the loggerhead turtle, *Caretta caretta*, in the wider Caribbean region. In: Marine Turtle Conservation in the Wider Caribbean Region: A Dialogue for Effective Regional Management, p. 36-40. Eckert, K.L., Abreu Grobois, F.A., Eds, St. Croix, U.S. Virgin Is, WIDECAST.

Godley, B., Broderick, A., Blackwood, S., Collins, L., Glover, K., McAldowie, C., McCulloch, D., McLeod, J. (1993): Survey of marine turtles nesting in Trinidad and Tobago. Mar. Turtle Newsl. **61**: 15-18.

Gorman, G.C., Licht, P., Dessauer, H.C., Boos, J.O. (1971): Reproductive failure among the hybridizing *Anolis* lizards of Trinidad. Syst. Zool. **20**: 1-18.

Hailey, A., Quesnel, V.C., Boos, H.E.A. (2009): The persistence of *Anolis trinitatis* as a naturalized lizard in Trinidad against hybridization pressure with *Anolis aeneus*. Appl. Herpetol. **6**: 275-294.

Hailey, A., Sookoo, N., Hernandez, E., Ramoo, D. (2007): The influence of density and ration level on cultured *Physalaemus pustulosus* tadpoles, and the mitigation of a crowding effect by soil substrate. Appl. Herpetol. **4**: 261-277.

Hailey, A., Sookoo, N., Mohammed, A., Khan, A. (2006): Factors affecting tadpole growth: Development of a rearing system for the Neotropical leptodactylid *Physalaemus pustulosus* for ecotoxicological studies. Appl. Herpetol. **3**: 111-128.

Hardy, J.D., Jr. (1982): Biogeography of Tobago, West Indies, with special reference to amphibians and reptiles. Bull. Maryland Herpetol. Soc. **18**: 37-142.

Hedges, S.B. (1993): Global amphibian declines: A perspective from the Caribbean. Biodivers. Conserv. **2**: 290-303.

Henderson, R.W., Berg, C.S. (2006, 2011): The herpetofauna of Grenada and the Grenada Grenadines: Conservation concerns. Appl. Herpetol. **3**: 197-213. Reprinted in: Conservation of Caribbean Island Herpetofaunas, vol. 2, p. 239-258. Hailey, A., Wilson, B.S., Horrocks, J.A., Eds, Leiden, Brill.

Huey, R.B., Deutsch, C.A., Tewksbury, J.J., Vitt, L.J., Hertz, P.E., Álvarez Pérez, H.J., Garland, T., Jr. (2009): Why tropical forest lizards are vulnerable to climate warming. Proc. R. Soc. B **276**: 1939-1948.

IUCN (2010): IUCN Red List of Threatened Species. Version 2010.1 (www.iucnredlist.org, accessed on 20 April 2010).

John, Y. (2009): Experts hunt for 'killer' snake. Trinidad & Tobago Express, 28 February 2009.

Jowers, M.J., Campbell-Palmer, R., Walsh, P.T., Downie, J.R. (2006): Intraspecific variation in the avoidance response of stream frog (*Mannophryne trinitatis*) tadpoles to fish and prawn predators. Herpetol. J. **16**: 337-346.

Jowers, M.J., Downie, J.R. (2004): Distribution of the frog *Mannophryne trinitatis* (Anura — Dendrobatidae) in Trinidad, West Indies. Living World, J. Trinidad Tobago Field Nat. Club, Suppl., 17-19.

Jowers, M.J., Downie, J.R. (2005): Tadpole deposition behaviour in male stream frogs *Mannophryne trinitatis* (Anura: Dendrobatidae). J. Nat. Hist. **39**: 3013-3027.

Jowers, M.J., Downie, J.R., Cohen, B.L. (2008): The golden tree frog of Trinidad, *Phyllodytes auratus* (Anura: Hylidae): systematic and conservation status. Stud. Neotrop. Fauna Envt. **43**: 181-188.

Kenny, J.S. (1969): The Amphibia of Trinidad. Studies on the Fauna of Curaçao and Other Caribbean Islands **29**: 1-78.

Kenny, J.S. (1977): The Amphibia of Trinidad — an addendum. Studies on the Fauna of Curaçao and Other Caribbean Islands **51**: 91-95.

Kenny, J.S. (1989): Hermatypic scleractinian corals of Trinidad. Studies on the Fauna of Curaçao and Other Caribbean Islands **123**: 83-100.

Kenny, J.S. (1995): Views From the Bridge: A Memoir on the Freshwater Fishes of Trinidad. Maracas, St. Joseph, Privately Printed.

Kenny, J.S. (2006): Heavy Industrialization in Trinidad — The Expansion into South-Western Peninsula and the Problems of Processes, Constitution, Law, National Policy, International Obligations and Carrying Capacity. Trinidad Aluminium Industry Symposium. San Fernando, South Trinidad Chamber of Industry and Commerce (http://www.energy.tt/symposium/Julien%20Kenny.pdf, accessed on 9 May 2010).

Kenny, J.S. (2008): The Biological Diversity of Trinidad and Tobago: A Naturalist's Notes. Port of Spain, Prospect press.

Knight, C.M., Gutzke, W.H.N., Quesnel, V.C. (2004): Shedding light on the luminous lizard (*Proctoporus shrevei*) of Trinidad: A brief natural history. Carib. J. Sci. **40**: 422-426.

Krebs, C.J. (1999): Ecological Methodology, 2nd edition. Menlo Park, CA, Benjamin/Cummings.

Lalsingh, G. (2008): Summary of Marine Turtle Nesting Activity 2008. Scarborough, Tobago, Save our Sea Turtles.

Lalsingh, G. (2009): Summary of Marine Turtle Nesting Activity 2009 — Turtle Beach, Grafton and Mt. Irvine Back Bay. Scarborough, Tobago, Save our Sea Turtles.

Lambie, I. (2005): The club's turtle project — how it all began. Q. Bull. Trinidad Tobago Fld Nat. Club December 2005: 1-2.

Lans, C., Harper, T., Georges, K., Bridgewater, E. (2001): Medicinal and ethnoveterinary remedies of hunters in Trinidad. BMC Compl. Alt. Med. **1**: 10.

Lee Lum, L. (2003): An Assessment of Incidental Turtle Catch in the Gillnet Fishery in Trinidad and Tobago, West Indies. Chaguaramas, Institute of Marine Affairs.

Lee Lum, L. (2005): Beach dynamics and nest distribution of the leatherback turtle (*Dermochelys coriacea*) at Grande Riviere Beach, Trinidad and Tobago. Rev. Biol. Trop. **53**: 239-248.

Lee Lum, L. (2006): Assessment of incidental sea turtle catch in the artisanal gillnet fishery in Trinidad and Tobago, West Indies. Appl. Herpetol. **3**: 357-368.

Lever, C. (1985): Naturalized Mammals of the World. London, Longman.

Library of Congress (2006): Country Studies: Trinidad and Tobago (http://countrystudies.us/, accessed on 18 March 2006).

Liddle, R.A. (1946): The Geology of Venezuela and Trinidad. Ithaca, N.Y., Paleontological Research Institution.

Livingstone, S.R. (2005): Report of olive ridley nesting on the north coast of Trinidad. Mar. Turtle Newsl. (109): 6-7.

Livingstone, S.R. (2006): Sea turtle ecology and conservation on the north coast of Trinidad. Ph.D. thesis, University of Glasgow.

Livingstone, S.R., Downie, J.R. (2005): Marine turtle conservation on the north coast of Trinidad — a Darwin Initiative project. BCG Testudo **6**: 3-16.

MacArthur, R.H. (1972): Geographical Ecology. New York, Harper and Row.

Manickchan, S.A. (2003): The Effect of Habitat Degradation and Biological Invasion on Species Richness and Abundance of Anuran Communities in Trinidad, West Indies. M.Phil. thesis, St. Augustine, Trinidad, The University of the West Indies.

Manzanilla, J., Jowers, M.J., La Marca, E., García-París, M. (2007): Taxonomic reassessment of *Mannophryne trinitatis* (Anura: Dendrobatidae) with description of a new species from Venezuela. Herpetol. J. **17**: 31-42.

Manzanilla, J., La Marca, E., García-París, M. (2009): Phylogenetic patterns of diversification in a clade of Neotropical frogs (Anura: Aromobatidae: *Mannophryne*). Biol. J. Linn. Soc. **97**: 185-199.

Marcovaldi, M.A. (2001): Status and distribution of the olive ridley turtle, *Lepidochelys olivacea*, in the wider Caribbean region. In: Marine Turtle Conservation in the Wider Caribbean Region, p. 52-56. Eckert, K.L., Abreu Grobois, F.A., Eds, St. Croix, U.S. Virgin Is, WIDECAST.

Mimura, N., Nurse, L., McLean, R.F., Agard, J., Briguglio, L., Lefale, P., Payet, R., Sem, G. (2007): Small islands. In: Climate Change 2007: Impacts, Adaptation and Vulnerability. Contribution of Working Group II to the Fourth Assessment Report of the Intergovernmental Panel on Climate Change, p. 687-716. Parry, M.L., Canziani, O.F., Palutikof, J.P., van der Linden, P.J., C.E. Hanson, C.E., Eds, Cambridge, UK, Cambridge University Press.

MLA (2004): Laws of Trinidad and Tobago. Revised edition, December 2004. Port of Spain, Ministry of Legal Affairs (http://rgd.legalaffairs.gov.tt/Laws2/main.html, accessed on 2 June 2010).

MPUE (2006): Draft National Action Programme to Combat Land Degradation in Trinidad and Tobago: 2006-2020. Port of Spain, Ministry of Public Utilities and the Environment.

Murphy, J.C. (1997): Amphibians and Reptiles of Trinidad and Tobago. Florida, Krieger.

Nathai-Gyan, N., James, C., Hislop, G. (1987): National Report for Trinidad and Tobago. Presented to the Western Atlantic Turtle Symposium II, Puerto Rico. Forestry Division, Ministry of Food Production, Marine Exploitation, Forestry and Environment.

Nellis, D.W., Everard, C.O.R. (1983): The biology of the mongoose in the Caribbean. Stud. Fauna Curaçao other Carib. Islands **64**: 1-162.

Ogilvy, V., Smith, J., Thornham, D. (2007): University of Glasgow Trinidad Expedition Report 2006. Glasgow, University of Glasgow.

Pemberton, C.A., Wilson, L.A., Garcia, G.W., Khan, A. (2002): Sustainable development of Caribbean agriculture. In: Natural Resource Management for Sustainable Development in the Caribbean, p. 277-306. Goodbody, I., Thomas-Hope, E., Eds, Kingston, Canoe Press of the University of the West Indies.

Peters, L. (2004): Community perspectives on marine turtle conservation in Trinidad. Living World, J. Trinidad Tobago Fld. Nat. Club, Suppl., 29.

Pounds, J.A., Bustamante, M.R., Coloma, L.A., Consuegra, J.A., Fogden, M.P.L., Foster, P.N., La Marca, E., Masters, K.L., Merino-Viteri, A., Puschendorf, R., Ron, S.R., Sánchez-Azofeifa, G.A., Still, C.J., Young, B.E. (2006): Widespread amphibian extinctions from epidemic disease driven by global warming. Nature **439**: 161-167.

Pritchard, P.C.H. (1984): Marine Turtles in Trinidad and Tobago: Report on a Consultancy to the F.A.O.

Rebel, T.P. (1974): Sea Turtles and the Turtle Industry of the West Indies, Florida, and the Gulf of Mexico. Coral Gables, FL, University of Miami Press.

Smith, J. (2007): University of Glasgow Exploration Society Trinidad Expedition Report 2007. Glasgow, University of Glasgow.

Spalding, M.D. (2004): A Guide to the Coral Reefs of the Caribbean. Berkeley, University of California Press.

Steadman, D.W., Jones, S. (2006): Long-term trends in prehistoric fishing and hunting on Tobago, West Indies. Latin Amer. Antiquity **17**: 316-334.

Steadman, D.W., Stokes, A.V. (2002): Changing exploitation of terrestrial vertebrates during the past 3000 years on Tobago, West Indies. Human Ecol. **30**: 339-367.

Uetz, P. (2010): The Reptile Database (http://www.reptile-database.org, accessed on 20 April 2010).

Urich, F.W. (1931): The mongoose in Trinidad. Trop. Agric. **8**: 95-97.

Van den Eynden, V. (2007): Matura National Park Environmentally Sensitive Area: Participatory Biological Baseline Survey. St. Augustine, Trinidad, The University of the West Indies.

Vitt, L.J., Avila-Pires, T.C.S., Caldwell, J.P., Oliveira, V.R.L. (1998): The impact of individual tree harvesting on thermal environment of lizards in Amazonian rain forest. Conserv. Biol. **12**: 654-664.

Watts, D. (1987): The West Indies: Patterns of Development, Culture and Environmental Change Since 1492. Cambridge, Cambridge University Press.

Waylen, K.A., McGowan, P.J.K., Pawi Study Group, Milner-Gulland, E.J. (2009): Ecotourism positively affects awareness and attitudes but not conservation behaviours: A case study at Grande Riviere, Trinidad. Oryx **43**: 343-351.

Wetterer, J.K., Davis, L.R., Jr. (2010): *Solenopsis invicta* (Hymenoptera: Formicidae) in the Lesser Antilles. Florida Entomol. **93**: 128-129.

Wetterer, J.K., Moore, J.A. (2005): Red imported fire ants (Hymenoptera: Formicidae) at gopher tortoise (Testudines: Testudinidae) burrows. Florida Entomol. **88**: 349-354.

Wetterer, J.K., Wood, L.D., Johnson, C., Krahe, H., Fitchett, S. (2007): Predaceous ants, beach replenishment, and nest placement by sea turtles. Environ. Entomol. **36**: 1084-1091.

White, G., Hailey, A. (2006): The establishment of *Anolis wattsi* as a naturalized exotic lizard in Trinidad. Appl. Herpetol. **3**: 11-26.

Williams, C.B. (1918): The food habits of the mongoose in Trinidad. Bull. Dept. Agric. Trinidad Tobago **17**: 167-186.

Williams, E.E. (1942): History of the People of Trinidad and Tobago. Reprinted 2002, New York, A&B Publishers.

Wilson, B.S., Horrocks, J.A., Hailey, A. (2006, 2011): Conservation of insular herpetofaunas in the West Indies. Appl. Herpetol. **3**: 181-195. Reprinted in: Conservation of Caribbean Island Herpetofaunas, vol. 1, p. 3-17. Hailey, A., Wilson, B.S., Horrocks, J.A., Eds, Leiden, Brill.

Wilson, S.M. (2007): The Archaeology of the Caribbean. New York, Cambridge University Press.

WTA (2010): World Travel Awards: Tobago Main Ridge Rain Forest (http://world-travel-awards.com/profile-2892-tobago-main-ridge-rain-forest, accessed on 4 June 2010).

Zimmerman, B.L., Bierregaard, R.O. (1986): Relevance of the equilibrium theory of island biogeography and species-area relations to conservation with a case from Amazonia. J. Biogeogr. **13**: 133-143.

Accepted: June 16, 2010 (JAH).

Index of genera and species